Mamura Abdumavlianova
Munira Sodikova
Mahbuba Mirsagatova

Recycled polymer materials

Mamura Abdumavlianova
Munira Sodikova
Mahbuba Mirsagatova

Recycled polymer materials

utilization, modification, recycling, application in the production of composite materials

ScienciaScripts

Imprint

Any brand names and product names mentioned in this book are subject to trademark, brand or patent protection and are trademarks or registered trademarks of their respective holders. The use of brand names, product names, common names, trade names, product descriptions etc. even without a particular marking in this work is in no way to be construed to mean that such names may be regarded as unrestricted in respect of trademark and brand protection legislation and could thus be used by anyone.

Cover image: www.ingimage.com

This book is a translation from the original published under ISBN 978-620-6-16403-6.

Publisher:
Sciencia Scripts
is a trademark of
Dodo Books Indian Ocean Ltd. and OmniScriptum S.R.L publishing group

120 High Road, East Finchley, London, N2 9ED, United Kingdom
Str. Armeneasca 28/1, office 1, Chisinau MD-2012, Republic of Moldova, Europe
Printed at: see last page
ISBN: 978-620-6-67675-1

Contents

INTRODUCTION

An important direction in the chemistry of high-molecular compounds is the creation of polymers and polymeric materials with new or improved properties. One of the ways to synthesise such polymers is modification of known and secondary polymers while maintaining the degree of polymerisation.

Convenient for modification is gaseous chlorine, which has been unclaimed until now, and which is formed in the technological chain of chemical production. Introduction of this chlorine into secondary polymer material creates conditions for obtaining ammonium polymers, which are widely used in various industries. It seems to be caused by the fact that the product obtained in this way is relatively new for the chemical industry of polymers. In this connection, the study of secondary polymer modification by chemical transformations of macromolecules and the study of physicochemical properties of modified polymer products is an important task.

Relevance of the work. Nowadays the problem of recycling of secondary polymeric materials is of topical importance, first of all from the point of view of environmental protection, but also with the fact that in conditions of shortage of polymeric raw materials plastic waste becomes a powerful raw material and energy resource.

Currently, there are several ways of utilisation of waste polymer materials, in particular burial in soil, incineration, thermal methods, creation of biodegradable polymers. But all these methods have significant disadvantages, namely: burial and incineration lead to environmental pollution and land reduction, thermal methods and creation of biodegradable polymers require significant financial costs, are technologically complex. Therefore, the most acceptable in terms of environmental protection and financial costs is the recycling of waste polymeric materials by chemical modification thereby creating polymer-polymer compositions with improved physicochemical and operational properties.

In this connection, the study of modification of polyethylene and its secondary product by chemical transformations, the study of physical and chemical properties of modified polymer products and the search for areas of their application is an urgent and important task.

Purpose of **work: The** purpose of the present work is chemical modification (chlorination, amination, etc.) of polyethylene and its wastes, study of properties of obtained new products.

The aim of the paper includes the following objectives:

- study of the process of halogenation of polyethylene and its wastes, determination of the optimal chlorination mode;
- Study of chemical transformation of chlorinated polyethylene modifications by amino derivative compounds;
- study of composition, structure and properties of synthesised products using modern methods of analysis;
- finding applications for the products obtained.

Practical significance of the work. Ammonium oligomers obtained on the basis of chlorinated polyethylene and amino compounds depending on the method and degree of chlorination, the nature of the amino compound can find wide application in various industries.

It is known that by adding a small amount of another polymer to a polymer composition the properties of the composition are sharply improved by a gamut of polymers with a wide range of properties. On the basis of chlorinated polyethylene it is possible to obtain polymer-polymer systems. Chlorinated polyethylene is used as an impact strength modifier in the production of rigid PVC products (window profile, panels, pipes), cables, can be used as an anti-corrosion coating for metal pipes. Properties of coatings and film-forming base of CPE can be changed within wide limits taking into account the operating conditions. Coatings based on CPE are weatherproof, resistant to most inorganic acids, alkalis and salts, they are elastic. Ammonium polymers are widely used in the production of hydrogels, can be used as flocculants for wastewater treatment and others.

CHAPTER 1

LITERATURE REVIEW

The experience of the most developed countries constantly reminds us of the well-known truth - the world does not stand still, reforming and democratisation of society, modernisation and renewal of the country is not a one-time, one-step process, but a continuous one.

Especially if we take into account that we live in the XXI century - the century of globalisation and the Internet, when the level of intellectual labour is of paramount importance, the scale and severity of competition in the world market are growing.

And in these conditions, first of all, those countries that do not rest on their laurels, soberly assess their resources and capabilities, and keep up with the rapidly changing world, win.

In this regard, the most important and priority task for us in the near future is further consistent deepening of structural reforms, continuous technological and technical renewal of production, introduction of modern information and communication systems.

Completely new industries and high-tech production facilities have been created literally from scratch, and their finished products now occupy a worthy place on the world market. We are talking about petrochemicals and the oil and gas industry,

It is through a seriously thought-out strategy of structural transformation, large-scale construction of new modern high-tech enterprises, modernisation and renovation of existing production that the country's GDP share will increase.

In accordance with the adopted Programme of priority measures to expand production volumes and master production of new types of competitive products, more than 270 investment projects will be implemented in 2020-2024.

In 2012, the implementation of projects that are extremely important for further diversification of our economy, including the start of construction of the Ustyurt gas chemical complex on the basis of the Surgil field, was envisaged and successfully released.

The third branch of the Central Asia-China gas pipeline with a total length of 1,830 kilometres and a design capacity of 25 billion cubic metres of gas per year was put into operation.The creation of special industrial zones - SIZ "Angren", SIZ "Navoi", SIZ "Jizzak" - became a powerful impetus for the development of high-tech industries.

As stated in the President's reports, the most important reserve and factor in economic growth and structural transformation of the economy is the deepening of the localisation of production and the expansion of inter-sectoral industrial cooperation.

The events taking place in the world today convincingly demonstrate that the creation of import-substituting industries on the basis of own raw materials ensures the economic independence of a country.

Our own experience of implementing localisation programmes shows the benefits we get from it, especially in terms of reducing dependence on external risks, creating domestic demand and saturating the market with necessary consumer goods and components, the fullest utilisation of the created production capacities, saving and rational use of foreign currency funds, and solving employment problems.

1.1. HALOGENATED POLYMERS

Speaking about the deepening of production localisation, expansion of inter-sectoral industrial cooperation and creation of import-substituting production facilities on the basis of own raw materials, one cannot but speak about production and processing of hydrocarbon products, in particular about chemical modification by introducing halogen atoms into the composition by halogenation. In fact, today, an important direction in the chemistry of high-molecular compounds is the creation of polymers and polymeric materials (from polymer waste) with new or improved properties.

According to the European Commission's report Waste Management in EU Member States, the UK is one of the best recyclers of plastic waste in Europe. Currently, the EU has a situation in which seven EU Member States prevent the landfilling of more than 90 per cent of plastics waste.

One of the ways of polymer waste recycling is synthesis of polymers from waste and modification of known polymers with preservation of degree of polymerisation. The most suitable for chemical modification are polymers whose macrochains contain unsaturated bonds, for example, carbon-carbon bonds. The activity of double $>C=C<$ bonds with respect to many reagents makes it possible to introduce substituents of different nature, including those containing heteroatoms, into the polymer chain. The latter can give the polymer such positive properties as, for example, increased heat resistance, hardness, impact resistance and others. Processing and production of polyethylene and utilisation of polyethylene products through modification of these products, use of modified products as plasticizers, fillers and other industries, occupies a significant place in the global industry.

Chlorinated polyethylene is one of the most widely produced products in the world, the world production capacity of this product is distributed among seven major producing companies and amounts to about 200 thousand tonnes.

Modification of polymers as a way to synthesise high molecular weight compounds with improved properties is of considerable interest from a practical and scientific point of view. One of the ways to modify the properties of polymers is the introduction of halogen atoms, most often chlorine, into macromolecules. Considering the scale of chlorine production, halogenation remains an economically justified method of modifying large-tonnage polymers. Halogenated polymers differ significantly from the original polymer products in their physicochemical properties, and the degree of change in physicochemical characteristics depends on the amount of halogen introduced into the polymer, which makes it possible to smoothly vary the properties within a wide range.

Chlorinated polyethylene is a white coloured powder used to improve the properties of PVC profiles, which are used in the manufacture of plastic windows, doors, and many other products. Products, in the production of which chlorinated polyethylene was used, have high weather resistance, colourability and low flammability.

Due to its physical and chemical properties, chlorinated polyethylene is a very valuable material for the production of cable and rubber products, as well as various types of seals.

This type of polymer material first began to be produced on an industrial scale in 1940, with over 200,000 tonnes of chlorinated polyethylene currently being produced worldwide.

Chlorinated polyethylene is produced from high density polyethylene by chlorination in aqueous medium. The chlorination process takes place in specialised tanks, which are covered with glass enamel. Polyethylene powder is mixed with water, chlorine gas is added to the resulting suspension, and the chlorination process is carried out according to a specific programme, which is most often protected by a trade patent.

After the chlorination process is completed, sulphuric acid is washed out of the resulting material, chlorinated polyethylene is mixed with certain additives and packaged in specialised containers, and the packaging markings indicate the percentage of chlorine in the product.

Currently, the production of chlorinated and chlorosulphated polyethylene is deployed in many countries. Many companies engaged in the production and sale of chlorinated and chlorosulphated polyethylene actively cooperate with various research institutes, which makes it possible to optimise the technological process and increase the quality and quantity of the product.

At present, the physicochemical properties of chlorosulfated polyethylene based on high-pressure polyethylene, low molecular weight polyethylene and secondary polyethylene have been studied. Swelling processes have been studied. The properties of chlorosulfured polyethylene and the influence of different hardeners and fillers on its properties are considered. In addition, the fields of application, compositions and properties of compositions based on the considered polymer are considered.

Traditionally, chlorinated polyethylene has been used to improve the impact strength of rigid polyvinyl chloride systems. Chlorinated polyethylene provides them with uniform mouldability properties as well as good impact resistance over a wide temperature range. These properties are most in demand in the production of PVC products such as profiles, sheets and pipes.Recently, the stand-alone use of CPE as an elastomer has become increasingly common. Mixed with fillers and various additives, CPE is well processed by calendering and extrusion. Products made of it are characterised by good weatherability, colourability and low flammability (roofing material). Chlorinated polyethylene can be cross-linked (vulcanised) by various methods. Cross-linked CPE (CHPE/TCPE) has excellent elasticity, thermal stability of shape, resistance to chemicals and ozone. These properties make CPE an extremely valuable raw material for the rubber and cable industries. These qualities also determine the prospects for the use of the material in the production of seals (cable sheaths, membranes, cushioning elements, absorption of acoustic shocks). The use of CPE mixed with polymers (blends) in the recycling of secondary plastic and rubber waste has great potential.

To date, the influence of flame retardants and highly dispersed fillers on flammability and physical and mechanical properties of recycled polyethylene has been studied. Recipe and technological parameters for obtaining fire-resistant tape on the basis of recycled polyethylene have been developed.

In most cases, chlorinated polyethylene is produced by chlorinating high density polyethylene (HDPE) in an aqueous medium. A water suspension is prepared from polyethylene powder with the addition of various additives, which is pumped into autoclaves coated with vitreous enamel, where - with a precise dosage of chlorine gas and a specific temperature programme - the chlorination process takes place. Hydrochloric acid is leached from the chlorinated polyethylene downloaded from the autoclave, then the product is dried, milled and, after mixing with certain additives, packaged.

As the chlorine content in polyethylene increases, its physical and mechanical properties change dramatically. When chlorinated, polyethylene gradually begins to lose its inherent crystallinity and becomes a highly elastic and rubber-like polymer with properties similar to polyvinyl chloride containing a large amount of plasticiser. As the chlorine content increases and the degree of crystallinity of the polymer decreases, its elasticity increases, reaching a maximum at 15-20% chlorine content, while the strength of the polymer decreases. The minimum strength of chlorinated polyethylene corresponds to 35-38% chlorine content.

Partially chlorinated polyethylene successfully replaces highly plasticised polyvinyl chloride. Such polyethylene is used for the production of films and artificial leather. Films of partially chlorinated polyethylene are superior in elasticity and oxidation resistance to polyethylene films. When heated above 190-200° C, gradual degradation of chlorinated polyethylene is observed.

By chlorination of polyolefins, products can be obtained that differ significantly in properties depending on the depth of reaction and the method of reaction. Even more these products differ from the initial polyolefin. For example, the authors have carried out chlorination of polydienes with high content of vinyl links by addition of dichlorocarbene on double bonds in a two-phase system with the use of interphase transfer catalysts. Chlorinated polydienes containing from 2.2 to 30% (wt.) chlorine were obtained. Vulcanisates were obtained and their mechanical properties were investigated.

In recent years, many researchers in the field of chemistry have been conducting research in the field of halogenation of polyolefins. For example, heterophase chlorination reactions of isobutylene have been studied. It was found that it is reasonable to carry out the reaction in the medium of target 3-chloro-2-methylpropane at boiling temperature of the reaction mixture and at molar ratio of reagents isobutylene:chlorine 1,0:1. In this case, the yield of 3-chloro 2-methylpropene increases from 86.7 to 63.9%, and the yield of each of the by-products is reduced by more than 40%. For the first time the process of obtaining composite olefin by reaction of copolymerisation of a-methylstyrene with dichloroisobutelenes contained in cube wastes of 3-chloro 2-methylpropene production has been investigated. It was found that the process should be carried out at a temperature of 90-100° C, the amount of catalyst 1% wt. and reaction time of 8 hours.

New polymeric halide-containing products based on syndiotactic 1,2-polybutadiene and oligomeric 1,2- polybutadiene have been synthesised by direct haloylation or catalytic hydrohalogenation. The effects of the degree of functionalisation of 1,2-polybutadiene on the characteristic viscosity of solutions, fluidity of polymer melts, thermal stability, thermomechanical, viscoelastic, and adhesion properties of modified polymer products have been studied.

As a thermoplastic material in its own right, chlorinated polyethylene, obtained even from high-density polyethylene, is relatively little used. This is due to lower rigidity, high deformability, impossibility to obtain a material with a high degree of filling, high cost, etc. than, for example, PVC. Mainly, CPE is used in blends with other plastics - PVC, polyolefins, ABS, etc. Blending of CPE with these plastics allows to obtain self-extinguishing materials with high impact resistance, frost resistance, etc. Graft or graft copolymerisation of CPE with vinyl monomers produces thermoplastic materials with good physical-mechanical and dielectric properties.

Thus, for example, chlorine derivatives of low molecular weight 1,2-polybutadiene are proposed as a component of PVC compositions improving conditions of polymer material processing. On the basis of halide-containing 1,2-polybutadiene halide adhesive compositions have been developed for use in the production of sticky polyvinylchloride films for insulation of pipelines. The developed adhesive compositions are submitted for patenting.

Chlorinated polyethylene (CPE) is of industrial importance, containing from 20 to 72% chlorine. The chlorine content and its distribution in the polymer determine the properties of the final product and depend on the method of production.

Chlorine content in industrial products is from 5 to 70%; thermoplastics (up to 14% chlorine), elastoplastics (15-23% chlorine), elastomers (2445%), rigid skin-like polymers (46-58%), glassy polymers (59-70% chlorine). Depending on the method and degree of chlorination, a gamut of polymers with a wide range of properties is obtained. With increasing the degree of chlorination up to 24-40%, the softening temperature first falls (up to 25500C), then, with further increase in chlorine content it increases, while the stiffness of the polymer increases.

1.2. BASIC METHODS OF HALOGENATION OF POLYOLEFINS

Chlorination by the action of molecular chlorine proceeds as a chain radical reaction:

$$Cl_2 \rightarrow Cl + Cl$$
$$RH + Cl \rightarrow R + HCl$$
$$R + Cl_2 \rightarrow RCl + Cl$$
$$R + SO_2 \rightarrow RSO_2$$
$$RSO_2 \rightarrow RSO_2Cl + Cl$$

etc.

Halogenated polymers are usually prepared according to a "conventional" *radical* polymerisation process, involving a halogenated monomer and a radical-generating agent that initiates the growth of polymer chains. *According to* this method, *the* time required for *the* polymer chain to grow until it reaches its final size is very short, often less *than* one second. Also, during the duration of polymerisation, new chains are born, grow and "die" essentially by direct recombination, by disproportionation or by reaction centre transfer. This "dying" *is of a* statistical, random nature; it depends, *in* particular, *on* the polymerisation temperature and on the viscosity of the medium. Thus, at the end of polymerisation, *a* halogenated polymer is obtained *which* has a random distribution of molecular weights having a polydispersity coefficient and *an* M_z/M_w ratio having values generally equal to *at* least two. Such a method does not enable the synthesis of polymers having a narrow and controlled molecular weight distribution. It also does not allow for the production of block copolymers.

It is generally agreed that *the* halogenated polymers obtained by the above process *are* valuable because they exhibit several advantageous properties, *in* particular high chemical resistance to alkalis and to alcohols and strong inhibition of flame spread. More specifically, the vinyl chloride polymers obtained by the above process are valuable because they can, if necessary, be made flexible by blending with plasticisers to produce flexible articles such as artificial leather.

However, they exhibit a number of weaknesses, the most important of which is that the numerous properties of these polymers that depend on molecular weight are heterogeneous on a molecular scale, and this is because these properties can change from one polymer chain to another in the same way as the molecular weight itself. In particular, the presence of very low molecular weight polymer chains in halogenated polymers of the prior art substantially reduces thermal stability; it also substantially impairs the mechanical properties of articles based on such polymers, in particular their tensile strength, abrasion resistance and scratch resistance. On the contrary, the presence of polymer chains of very high molecular weight in these same halogenated polymers makes their processing more difficult (higher melt viscosity); it also significantly impairs the mechanical properties of plasticised articles made from them, due to the lack of gelation (imperfect mixing, at the molecular level, of the halogenated polymer and the plasticiser) and other properties of these articles, such as transparency and final surface finish.

However, the polydispersity coefficient and the M_z/M_w ratio of the halogenated polymers synthesised by this process, although lower than these parameters for the halogenated polymers obtained by the conventional process, remain high (for example, the polydispersity coefficient, for vinyl chloride homopolymers, is equal to or greater than 1.70), and this shows that the process under consideration is strongly influenced by side reactions in the same way as the conventional process, chain transfer reactions and/or disproportionation and/or recombination reactions, which are prone to occur in the halogenated polymers obtained by the conventional process. Another disadvantage of this method, known from the literature, is the need to use an iodinated organic chain transfer agent, which is generally expensive and increases the cost of the halogenated polymer.

The term "halogenated polymers" is understood to mean both homopolymers of halogenated monomers and copolymers which the halogenated monomers form with each other or with non-halogenated monomers. These copolymers may in particular be random copolymers, block copolymers or grafted copolymers.

The term "halogenated monomer" is understood to mean any ethylenically unsaturated monomer that contains at least one halogen atom. Examples of halogenated monomers may include halogenated vinyl monomers, halogenated styrene monomers such as 4-bromostyrene, halogenated (meth)acrylic monomers such as trifluoroethylacrylate, and halogenated conjugated dienes such as chloroprene.

The development of a method for the radical polymerisation of halogenated monomers which is more efficient than methods known from the literature appears to the person skilled in the art to be a particularly difficult problem, since halogenated monomers, in particular vinyl chloride, vinyl   fluoride,   vinylidene chloride and vinylidene fluoride, are regarded as inherently prone to transfer the radical reaction centre of growing polymer chains to themselves (chain transfer reactions to monomer) or to inactive polymer chains (chain transfer reactions to polymer).

The first chlorinated polyethylene was produced in 1936. Then gaseous chlorine was passed through solutions or suspensions of low-density polyethylene in carbon tetrachloride or acetic acid in the presence of catalysts: aluminium chloride, iron chloride, titanium chloride. In addition to carbon tetrachloride, water, tetrachloroethane, chloroform were used as reaction media for PE chlorination. It was possible to treat PE under pressure with liquid chlorine, in which it dissolves in the process of chlorination.

The main methods of production of chlorinated polyethylene are based on the reaction of fine powder of solid polyethylene with chlorine in liquid or gaseous medium - chlorination "in suspension", or on the reaction of polyethylene with chlorine in the solution of chlorine-resistant solvent - chlorination "in solution". Chlorination of polyethylene occurs under the action of radical initiators (peroxides and azo compounds), photo- and ionising radiation.

Typically, chlorination of polyethylene is carried out in an organic solvent or in an aqueous suspension. Methods such as chlorination of solid polyethylene in a fluidised bed, chlorination in an emulsion or in a melt are also known. However, these methods are not of widespread industrial importance.

It is known method of CPE production in mass, which consists in chlorination of solid, finely dispersed PE particles by a flow of gaseous chlorine in the presence of radical initiator. The process is strongly exothermic, which leads to local overheating, causing melting and clumping of particles and even destruction of the polymer, so the chlorinated mass is heated in stages, bringing it to 1320C. Step-by-step heating somewhat reduces the possibility of local overheating, however, does not sufficiently ensure a uniform distribution of the desired properties in the resulting CPE, which, in particular, contains a significant amount of brittle and rigid PE. These disadvantages of the known method limit its wide application.

We have studied the most frequently used methods of chlorination of polyolefins. For example, chlorination of polymers and copolymers of olefinic hydrocarbons is carried out in hexachloroacetone medium in the presence of initiator. The precipitation is carried out by adding an aliphatic hydrocarbon to the reaction mixture. After separation of the product from the reaction mixture by decantation or filtration, it is washed with precipitant, then treated with water under intensive stirring at a temperature of 20-100°C. Such treatment allows removing residual amounts of hexachloroacetone from the product, eliminating the adhesion of the product particles and thus obtaining it in a convenient form for use (powder or small granules). These products are widely used in the manufacture of chemically and weather-resistant varnishes, enamels, adhesives, films, electrical insulating materials, as an impact resistance modifier for PVC-based compositions, and are especially widely used in the rubber and tyre industries.

At room temperature polyethylene is insoluble in common solvents. At temperatures above 60^0 C, depending on the degree of crystallinity, it dissolves in chlorinated aliphatic and aromatic hydrocarbons. The most common solvent used is carbon tetrachloride, which is a very effective solvent for low density polyethylene near its boiling point (76^0 C). Higher temperatures ($80-110^0$ C) and pressures are required for chlorination of high-crystalline (low-pressure) polyethylene.

In industrial production two methods are used to obtain chlorinated polyethylene - gaseous chlorination of polymer solution in chlorinated solvent or chlorination of polymer suspension. The polymer is separated from solutions by distillation of solvent with steam in the form of azeotropic mixture, from suspension by filtration or centrifugation.

The known method of obtaining CPE in solution consists in treating PE with gaseous chlorine in carbon tetrachloride medium at atmospheric pressure, at the boiling point of carbon tetrachloride and in the presence of a substance initiator.

The advantage of the considered method of obtaining CPE, performed in solution, is that the obtained product is characterised by homogeneous properties, the process of PE chlorination can be easily regulated. However, the main disadvantage of the known method is a small solubility of PE in solvents inert from the point of view of chlorination, in particular in carbon tetrachloride, which causes their high consumption. In addition, significant difficulties are caused by the removal of the chlorinated product from the solution and the need to grind the sticking particles of CPE after the removal of solvent. Solvent regeneration requires special equipment and complicates the technology of implementation of the method.

A method of producing chlorinated polyethylene by passing chlorine gas through a suspension of low-pressure polyethylene under heating, characterised in that an organic compound selected from the group consisting of C [-'.ide n 8 12;m 18

21,$X^2\text{C1})_nx$,where X F or C1;n 17,$H^2)nX$,where X C1 or CH2C1;n 6 8,$(-CF2CF2-)n$,where n 6 12, at a polyethylene content of not more than 25 wt%.The choice of fluorinated liquids as a suspending medium for obtaining CPE was determined taking into account their properties: fluorinated liquids are hydrophobic, which helps to prevent sticking of PE particles during chlorination; fluorinated liquids do not dissolve hydrocarbons, including PE and CPE; fluorinated liquids do not interact with chlorine in the process of obtaining CPE, so they can be used repeatedly without purification. Fluorinated liquids with the above mentioned coefficients correspond to the boiling point, which allows active chlorination at optimal temperatures, and also correspond to the optimal viscosity, which ensures the easiest removal of traces of fluorinated liquids from CPE when drying or centrifugation. The found features of the method implementation do not require the PE chlorination process to be carried out at elevated pressure. Furthermore, it was confirmed that the claimed process can be carried out in a recycling mode. The slurry composition is inert with respect to the equipment in which the PE chlorination is carried out, so that the danger of its corrosion is excluded. The claimed method allows to easily adjust the degree of chlorination and other physical and mechanical properties of the obtained product by using PE particles in a wide range of sizes, time and temperature of chlorination.

In the method of separation of chlorosulfated or chlorinated polyethylene by application of their solutions in carbon tetrachloride on a moving surface, which is a drum heated from inside, differing in that the surface of the drum is covered with fluorinated polymer, and application of the solution in which 3.6-4.0 wt.% of water is dispersed. of water, is carried out on the upper part of the drum in the course of rotation with subsequent distribution of the solution over the surface by means of a distributing device and the film is removed from the surface heated up to 80 100° C with the content of less than 0.3 wt. of carbon tetrachloride.

In the method of production of chlorinated polystyrene the set goal is achieved by the fact that as a chlorinating agent a mixture of aqueous solution of sodium hypochlorite and hydrochloric acid is used in the ratio from 1:1 to 1.5:1.The proposed method is carried out at room temperature and in the absence of any catalysts. It allows to simplify considerably the technological scheme of the process and to exclude additional means and complex devices necessary for cooling or condensation of the reaction mixture and formed secondary products, etc.Application of the mixture of aqueous solution of sodium hypochlorite and hydrochloric acid (HCl) as a chlorinating agent instead of gaseous chlorine creates a number of conveniences; HC1 is not emitted, for collection of which additional technological scheme and related equipment and means are necessary, the used chlorinating mixture in comparison with Also, the method of chlorination of high density polyethylene up to the chlorine content of 15-20 weight.% with subsequent dochlorination up to the chlorine content in the final product of 35-45% at 85-130 C and pressure of 1-10 atm. in the presence of initiator, and the process is carried out in the medium of carbon tetrachloride.The disadvantage of the known method is that the duration of the process is 3-5 hours.The purpose of the invention is to intensify the process. This purpose is achieved by the fact that in the method of chlorination of high density polyethylene to the chlorine content of 15-20 weight.% with subsequent pre-chlorination at 85-130° C and 110 atm. in the presence of an initiator, chlorination to the chlorine content of 15-20 weight.% is carried out in the presence of water taken in the amount of 10-100 % of the weight of polyethylene, and pre-chlorination is carried out to the chlorine content in the final product of 30-51.5 weight.%.

Obtaining chlorine-containing frost-resistant polymer with high physical and mechanical properties is achieved by using divinylstyrene thermoplastic elastoplast as an initial polymer. Chlorinated divinylstyrene thermoplastic elastoplast according to the proposed method is obtained by chlorine treatment of divinylstyrene thermoplastic elastoplast in carbon tetrachloride or benzene at a temperature not exceeding 50° C. The chlorine-containing thermoplastic elastoplast is precipitated from the solution, dried and then analysed. The obtained chlorine-containing tsrmoelastoplast is precipitated out of the solution, dried and then analysed. Chlorine content in the samples is determined analytically after re-precipitation from methyl or isopropyl alcohol.

A method for halogenation of elastomers has been developed which so improves the dispersion and diffusion of the halogen that high local concentrations of said halogen can be avoided. The invention furthermore relates to a device for carrying out the above method.A method for halogenating elastomers in which a halogenating agent is introduced into a solution of an elastomer in an organic solvent, essentially characterised in that said halogenating agent is mixed with a continuous stream of said elastomer solution with said halogenating agent eventually dissolving and interacting with said elastomer in said continuous stream, said continuous stream of elastomer solution being maintained in a turbulence In an apparatus for carrying out the above method, the turbulence of the solution is maintained by mechanical means that promote turbulence, wherein the means may be static or dynamic, such as Raschig rings loaded in a tube, such as a column type reactor, in which the organic elastomer solution is forced to flow continuously. Advantageously, an inert gas is added to said elastomer solution in order to improve halogen dispersion, increase turbulence and reduce the local concentrations of the halogenating agent and the resulting hydrogen halide. The addition of the halogenating agent to the continuous flow of said elastomer solution is carried out at atmospheric pressure, a defined high reaction rate under the operating conditions used.

A method of halogenating elastomers, comprising supplying a halogenating agent with an excess of inert gas or a solution thereof to a continuous stream of unsaturated elastomer in an organic solvent, which are mixed and subjected to interaction under conditions of turbulent motion without flow inversion phenomenon, characterised in that the gas-liquid or liquid mixture is periodically separated on hollow nozzles into mutually communicating peripheral and axial flows, which are periodically combined between the nozzles in porous separation layers and finally combined after the last nozzle along the flow path. Device for halogenation of elastomers, including a cylindrical body with hollow nozzles located in it, reagent input and output fittings, differing in that inside the body sequentially along the length of the reactor placed flow splitter and nozzles, and nozzles on the outer surface have grooves that form a helical channel or channels with the inner surface of the body wall, and the inner surface of the nozzles at the ends is made with conical ledges with a diameter 1,3-3 times smaller than in the cylindrical part of the inner cavity of the nozzle, and the helical channel or channels and the inner cavity of the nozzle are connected by holes and/or slots, and behind each nozzle there is a porous separating layer having a height not more than 1/2 of the nozzle height.

A method of halogenation of elastomers, comprising supplying a halogenating agent with an excess of inert gas or a solution thereof to a continuous flow of unsaturated elastomer in an organic solvent, which are mixed and subjected to interaction under conditions of turbulent motion without flow inversion, distinguished by the fact that the resulting gas-liquid or liquid mixture is separated into at least two streams - peripheral, moving in a spiral direction, and axial - in nozzles installed sequentially along the flow, and the mutually communicating streams are combined into a single stream at the outlet of the last nozzle. The device for halogenation of elastomers, including a cylindrical body with hollow nozzles located in it in sequence along the length of the body, nozzles input elastomer solution, gas or gas mixture and nozzle outlet of gas-liquid or liquid flow, differing in that the outer surface of the nozzles grooves, forming a helical channel or channels with the inner surface of the body wall, and the inner surface of the nozzles at the ends is made with conical ledges with a diameter 1.3-3 times smaller than in the cylindrical part of the inner cavity of the nozzle, and the helical channel or channels and the inner cavity of the nozzles are connected by holes and/or slots. The device differs in that one or more static mixer, such as an Archimedes spiral, is installed behind the last nozzle along the course of the gas-liquid or liquid flow.

The method of haloydising butyl rubber allows a substantial increase in the conversion of halogens compared to previously known methods.Continuous method of haloydising butyl rubber, including introduction of chlorine or bromine into butyl rubber solution in a solvent inert with respect to them, mixing of chlorine or bromine with a continuous flow of the said polymer solution at dissolution of halogen and its interaction with butyl rubber at turbulent movement of the solution, followed by separation of the gas-liquid flow and neutralisation, differing in that the gas stream formed during separation is directed to an additional stage for interaction with the initial butyl rubber solution before its neutralisation.

A method for obtaining brominated butyl rubber has been developed and can be used in petrochemical and chemical industries. Bromination of butyl rubber includes treatment of 10-25% solution of butyl rubber in C5-C8 hydrocarbon solvent with bromine water followed by neutralisation of reaction mass, introduction of stabilisers and anti-agglomerator, water degassing and drying of bromobutyl rubber. Bromine water is obtained by dissolving bromine in water or by electrolysis of an aqueous solution containing bromide ions. To neutralise excess bromine in the bromobutyl rubber solution, organic acid and/or alkaline or alkaline earth metal sulphite are used. Neutralisation is carried out in one step - by treating the reaction mass with calcium or sodium sulphite or in two steps, whereby at the first step the reaction mass is pre-treated with organic acid, and at the second step sodium sulphite is introduced into the reaction mass.

There is a known method of obtaining halogenated polymers by interaction of halogen with polymer, differing in that the crumb of polymer or its solution in non-aqueous medium is suspended in an aqueous solution of halogen-containing compounds in a reactor and the interaction of polymer with halogen is carried out at the expense of halogen electrochemically generated inside the reaction space of the reactor where halogenation is carried out.

A halogenated or halogensulfated (co)butene-1 polymer obtained in tetrachloroethane or chlorobenzene solution and containing chemically bound chlorine, and/or bromine, or chlorine and sulfur, or bromine and sulfur, or simultaneously chlorine, bromine and sulfur in the form of SO_2a- and SO_2Br-groups, wherein the content of bound chlorine and/or bromine is 1-73 wt.%, sulfur 0.24.0 wt.%. A binder for an adhesive adhesive or paint composition, which is a modified halogenated or halogensulfated (co)butene-1 polymer obtained in a solution of chlorobenzene or tetrachloroethane and containing chemically bound chlorine and/or bromine and sulfur in an amount of 58-73 wt.% chlorine or 56-70 wt.% bromine and sulfur, respectively.wt.% chlorine or 56-70 wt.% bromine, or 56-70 wt.% chlorine and 0.3-4.0 wt.% sulfur, or 58-70 wt.% bromine and 0.2-3.76 wt.% sulfur, or simultaneously chlorine and bromine at a total content of 58-70 wt.% thereof, or chlorine and bromine at a total content of 56-73 wt.% thereof and sulfur in an amount of 0.2-4.0 wt.% in the form of SO_2CI- and SO_2Br-groups.

In the method of obtaining perchlorovinyl resin by treatment with chlorine gas of a solution of polyvinyl chloride resin in an organochlorine solvent in the presence of an initiator with subsequent separation of resin, it is distinguished by the fact that treatment with chlorine gas is carried out in the presence of a 1% solution of dicetyl peroxydicarbonate in dichloroethane as an initiator at 60-105°C for 4.5 hours. with flow rate of gaseous chlorine 60-90 m3/h, with dichloroethane and trichloroethane used as an organochlorine solvent at their mass ratio of 1,0:0,1.

Chlorination of polymers and copolymers of vinyl chloride is carried out with chlorine gas in hexachloroacetone at elevated temperature, usually in the presence of an initiator. The target product is precipitated by treating the reaction mixture with aliphatic or cycloaliphatic hydrocarbons. Prior to precipitation of the target product, the reaction mixture is treated with a diluent. The target product is then separated by filtration, washed and dried. As diluents, solvents capable of dissolving chlorinated polymers and copolymers of vinyl chloride, preferably selected from the group consisting of aliphatic and aromatic chlorohydrocarbons, ketones, alkyl acetates, aromatic hydrocarbons or various mixtures thereof, are used. Aliphatic or cycloaliphatic hydrocarbons containing 3-12 carbon atoms in their structure or various mixtures thereof are used as precipitants. The technical result is simplification of solvent regeneration, elimination of the use of toxic methanol and formation of wastewater containing organochlorine impurities. The use of hexachloroacetone as a solvent eliminates the formation of by-products of chlorination of solvents, and therefore solves the problem of utilisation and destruction of these wastes.

In a method of obtaining chlorinated polyvinyl chloride by treatment with chlorine gas of a solution of vinyl chloride polymer in chlorobenzene in the presence of azobisisobutyronitrile at 100-105°C with subsequent separation, a copolymer of vinyl chloride with 1.5 4.0 wt% of a monoester based on maleic anhydride and oxyethylated alcohol is used as a polymer for chlorination. The invention relates to methods of obtaining highly adhesive chlorinated polyvinyl chloride (CPVC), which can be widely used in the production of chemically resistant varnishes, paints and enamels.CPVC should be referred to polymers characterised by extremely low adhesion.

In the method for the preparation of chlorinated polydienes, the degree of transformation of alkali metal hydroxide can be increased, which provides a more efficient use of the reagents used. In addition, the preparation of chlorinated polydienes does not require the use of hydrocarbon solvent or 20 mol excess of chloroform, as well as the preliminary dissolution of KMP in chloroform, which simplifies the process.The method allows to obtain chlorinated polydienes with different chlorine content (up to 35 wt.%), while the molecular weight characteristics of chlorinated polydienes do not undergo significant changes.

A method of obtaining chlorinated polydienes by interaction of polydienes with chloroform in the presence of an interfacial transfer catalyst and an aqueous solution of alkali metal hydroxide at its fractional feeding, consisting in that the interaction is carried out by feeding the entire calculated amount of chloroform directly into the polymer, and the interfacial transfer catalyst is fed into the polymer solution in pure form and the process is carried out at a molar ratio of polydiene: chloroform: alkali metal hydroxide: interfacial transfer catalyst 1:(2.0- 3.0):(0.1-2.5):(0.001-0.0018).

The chlorination process is described by a flowchart:

$$\sim(CH_2-CH=CH-CH_2)_n\sim \ + \ CHCl_3 + \ MOH \ \xrightarrow{\text{КМП}} \ \sim(CH_2-CH-CH-CH_2)_n\sim$$
$$\underset{Cl \quad Cl}{\overset{\displaystyle C}{\diagup\diagdown}}$$
$$+ \ MCl + H_2O$$

where M is sodium or potassium,

KMP is a catalyst for interfacial transfer.

An indicator of the efficiency of the process for obtaining modified polydiene is the degree of conversion of alkali metal hydroxide. The degree of transformation of alkali metal hydroxide is from 30 to 40%.

In general, chlorination in aqueous suspension became widespread after the appearance of high-density polyethylene. A 1-20% solution of methyl sulfoxide is often used as an additive to ensure swelling of the polymer and uniformity of chlorination. To carry out the process, suspensions containing from 5 to 20% of polyethylene suspended in water, in aqueous 6-8 n. HCl solution, in concentrated II2SO)$_4$ or 0.5-2^1- aqueous CaCl solution$_2$ are used. At the same time it is ensured that the suspension is resistant to agglomeration and foam formation during the reaction. Up to 2% of surfactants are introduced into the reaction mixture. Quaternary ammonium salts are added to the reaction mixture to prevent the formation of electrostatic charges.

High-density polyethylene is most suitable for chlorination in the solid phase, while low-density polyethylene must be converted into a swollen state by special treatment. As a chlorinating agent gaseous chlorine or a mixture of chlorine and nitrogen is used. If the mixture contains up to 10% chlorine, mainly amorphous parts of high density PE are subjected to chlorination, and the chlorinated product contains non-chlorinated polyethylene crystallites. If the mixture contains more than 10% chlorine, the resulting product generally has an amorphous structure.

Chlorination in the fluidised bed is associated with more stringent requirements to the technology. Usually PE with a high degree of pulverisation (particle size 5.0-20.0 μm) in suspension is irradiated with UV- and U-rays. The temperature is raised to 150-250^0 C to accelerate the process. Chlorination of PE in fluidised bed produces a product with 40-50% chlorine content, characterised by high homogeneity, heat and light stability and other better properties compared to the product obtained by chlorination in solid phase.

Special methods of chlorination of polyethylene, which are not widely used in technology, include chlorination with liquid chlorine, chlorination of polymers in mass (films, fibres), as well as combined chlorination (at the first stage - in a fluidised bed, and at the second - in an aqueous suspension or organic solvent).

Chlorination in liquid chlorine is carried out at temperature 20-500C under pressure or at temperature - 350C at atmospheric pressure in the presence of benzoyl peroxide or under the action of UV radiation and intensive mechanical stirring. The resulting product, depending on the duration of the process, may contain up to 69% chlorine.

The processes of chlorination of oligo- and polydienes in the liquid phase at moderate temperatures without external initiation have been studied. The influence of temperature and depth of chlorination on the ratio of substituent and attachment chlorination was revealed, and the possibilities of obtaining polymeric materials were revealed.

1.3. MODIFICATION OF HALOGENATED HYDROCARBONS

It is known that when low molecular weight chloroalkyls interact with amines and ammonia, amine derivatives are formed relatively easily. The reaction with amines is also very characteristic for chlorinated polymers.

The method of modification of thermoplastic polymers by photochemical chlorination is known.The proposed method differs from the known method in that thermoplastic polymers after photochemical chlorination are treated with ammonia.The essence of the proposed method is to give semiconducting properties only to the surface of the polymer material without changing the physical and mechanical properties of its volume.This is achieved as follows.Polymers in the form of films, fibres or finished products are irradiated with light in an atmosphere of chlorine. Since the chlorine is mainly in contact with the polymer surface, the result is photochemical chlorination of the surface. By varying the temperature at which the illumination is performed, the duration of irradiation, and the concentration of chlorine in the reaction vessel, chlorination can be carried out to varying depths, of the polymer, and to varying degrees. The chlorinated polymer outside the chlorine environment is then treated with ammonia, dehydrochlorination of polymers and simultaneous formation of conjugated double bonds, which give polymers semiconducting properties, Chlorination and dehydrochlorination occur in the surface layer of the polymer, the thickness of which is less than the thickness of the sample. Mechanical processing to give the necessary shape and dimensions is carried out before applying the semiconductor layer by the opiton method.Method of modification of thermoplastic polymers, for example polyethylene, or polypropylene, or polyvinylchloride, by photo-chemical chlorination, differing in that, in order to improve physical and mechanical properties at elevated temperatures and to give semiconductor properties, the above polymers after chlorination are treated with ammonia.

Studying the interaction of CPE with aniline, di-n-butylamine and ammonia, Krenzel et al. concluded that the main direction of the reaction is dehydrochlorination with the formation of double bonds in the polymer chains, amination and intermolecular imination.

As the chlorine content in the polymer increases, its tendency to dehydrochlorination reaction with the formation of C=C bonds increases. This is manifested in the deepening of the colour of the formed product. The share of nitrogen attached to CPE is small and does not exceed 2%.

Later it was established that the amination of CPE occurs as a result of substitution of chlorine atoms at tertiary carbon atoms, since the content of attached nitrogen corresponds to the number of branching in the polymer chains. According to IR spectroscopy data, the most significant changes in the structure of CPE are observed during amination with primary amines - monoethanoamine and butylamine:

$$
\begin{array}{c}
R \\
| \\
....-CH_2-C-CH_2-...+H_2N-CH_2-CH_2-OH \rightarrow \\
| \\
Cl
\end{array}
$$

$$
\begin{array}{c}
R \\
| \\
\rightarrow HCl+...-CH_2-C\ -CH_2... \\
| \\
HN\ -CH_2\ -CH_2\ -OH
\end{array}
\qquad
\left[
\begin{array}{c}
R \\
| \\
...-CH_2-C-CH_2-... \\
| \\
HCl-\ HN-\ CH_2\ -CH_2-OH
\end{array}
\right]
$$

Transverse bonds in the polymer are formed during the subsequent heat treatment *(160⁰ C, 30* min.*) of the* amminated products. At the same time the reduction of bound chlorine content continues, *a* reduction of bound nitrogen is also observed, *the* polymer becomes insoluble*:*

$$
\begin{array}{ccc}
R & & H \\
| & & |\ \delta^+ \\
....-CH_2-C-CH_2-...+...-CH_2-C\ -\ CH_2-...\rightarrow \\
| & & |\ \delta^- \\
HN-R' & & Cl
\end{array}
$$

$$
\begin{array}{c}
R \\
| \\
\rightarrow R'NH_2+...-CH_2-\ C\ -CH_2-... \\
| \\
...-CH_2-\ C\ -CH_2-... \\
| \\
Cl
\end{array}
$$

During amination of CPE and further heat treatment, intramolecular hydrogen chloride detachment and formation of double bonds in polymer macromolecules also occur. Amination of CPE and subsequent heat treatment increases the polymer strength and improves its anti-corrosion properties, which confirms the formation of cross-linked structures.

When the surface of CPE is treated with ammonia for 2-3 hours, dehydrochlorination of the polymer occurs and simultaneous formation of conjugated double bonds in its surface layer. This technique is used to give semiconducting properties to the surface of polymer material.

Reactions of chlorinated polymers with oligomers obtained by the addition of hydrocarbon radicals on the amino group are of practical importance.

Chlorinated polymers react not only with amines but also with their salts.

The alkyl groups in quaternary organic ammonium compounds may also be substituted with functional groups such as hydroxyl, amine, amide, imine, carbonyl, carboxyl, silane, siloxane, simple ether, thioether, ester, ester complex, nitrile group, sulfonic acid residue, epoxy group, carboxylic acid anhydride fragment, carbonyl groups and/or halogen atoms such as fluorine (F) or chlorine (Cl).

Heating of chlorosulfured polyethylene with salts of hexamethylenediamine and adipic or sebacic acids (salts of AG and SG, respectively) is accompanied by crosslinking of the polymer.

Analysis of IR spectra of products of heating of CSPE with SG salt shows that salt consumption occurs at least as a result of two parallel processes: polycondensation of the salt and its reaction with the polymer. When the initial salt is replaced by the heated salt, the degree of crosslinking of the polymer decreases. If instead of the SG salt we introduce the product of its complete condensation (obtained by prolonged heating, in a sealed ampoule at 220^0 C), then no cross-linking is observed at all. Consequently, polycondensation is a side process, and direct reactions of CSPE with SG salt lead to crosslinking.

In CSPE macromolecules, both chlorosulfone groups and chlorine atoms can react with amines. However, when heated with CSPE SG salt, the main reaction is polycondensation.

Consequently, in the case of CSPE, it is the interaction of the SG salt with chlorosulfone groups that leads to crosslinking of the polymer. The first stage of the reaction is the formation of ionised pendants:

$$NH_3^+ - OOC$$

$$KaSO_2Cl + (CH_2)_6 \qquad (CH_2)_8 \rightarrow KaSO_2NH(CH_2)_6 NH_3^+ Cl^- + HOOC(CH_2)_8 COOH$$

$$NH_3^+ - OOC$$

With increasing duration of heating$^+$ NH3Cl NH3Cl$^-$ - groups are gradually consumed with the formation of secondary sulfonamides.

The possibility of reaction of SO2Cl-rpynn$^+$ NH3Cl$^-$ - groups is confirmed by the fact that when CSPE is heated with HMDA dihydrochloride, crosslinking of the polymer is not less intense than with the SG salt.

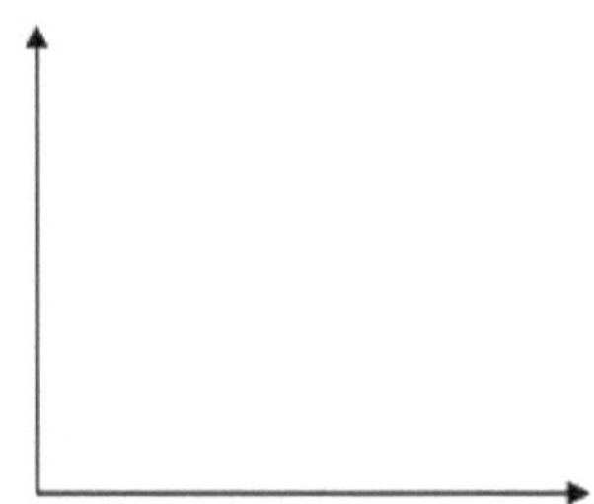

Fig. 1.1 - Crosslinking kinetics (based on equilibrium swelling) at 160° C of CPE (1-3 and CSPE (4) with a mixture of 15 wt.% of magnesium oxide and SG salt taken in amounts of 6 (1), 8 (2), 10 (3) and 6.5 (4) wt.%.

As can be seen from Fig. 1.1, at the same content of SG salt and magnesium oxide, the crosslinking of CPE proceeds at a lower rate, and the resulting

products have a significantly lower crosslinking limit than in the case of CSP.

The rate and extent of crosslinking of CPE by the mixture of SG and magnesium oxide increase significantly with the introduction of sulphur.

Thus , when CSPE interacts with hexamethylenediamine salts, pendants representing ionised sulphonamides are formed at the beginning, which then form disulphonimide cross-links. Since polar products are formed at all stages of the reaction, they retain sorption interactions with the original salt microparticle.

1.4. APPLICATION OF HALOGEN-CONTAINING POLYMERS

The main fields of application of chlorinated polyethylene are production of non-combustible, chemical-resistant flooring, production of artificial leather, foams, films, sheets, profile products, Turb, laminated materials, coatings for cables, wires, etc. CPE, as well as thermoplastic materials, can be applied by gas-flame spraying on products made of metals, textiles, glass, paper, wood.

So, for example, as a result of complex research of compositions on the basis of chlorinated polyolefins the possibility of significant increase of adhesion strength and other mechanical characteristics of protective coatings on the basis of chlorosulphated polyethylene due to curing by amines has been established. The optimal hardener for these coatings are aromatic amines, the effect of which is reduced to the formation of compounds such as sulfonamides forming strong bonds with the metal surface in the process of curing.

For the first time halide mechanochemical modification of diene and olefin elastomers, in which chlorine-containing limiting hydrocarbons capable of telomerisation and dehydrochlorination reactions were used as a halide-modifying component, the structures and properties of the obtained chlorine-containing elastomers were studied. The main technological parameters of obtaining chlorine-containing elastomers by means of halide mechanochemical modification and creation on their basis of heat-, ozone-, oil and petrol-resistant, etc. materials are developed. Materials.

A new protective polymer-textile material with enhanced protective properties (provides simultaneous resistance to aggressive substances and high gas tightness) has been developed. A multilayer polymer-textile material with a fluorine-containing outer layer, a butyl rubber inner layer, and a polymer composition based on a mixture of fluorine rubber of SKF-26 grade and fluoroplastic of F-62 grade is chosen as the object of research.

Compositions containing fluoropolymer as a co-setting agent (unsaturated hydrocarbon with perfluoro-aliphatic groups) are being developed. The presence of this group in the composition improves the removal of the product from the mould.

The authors of the work have studied the prospects for the use of fluoropolymers in the textile industry. It is proposed to apply fluorine-containing materials to fibrous materials either from true solutions or by introducing nano- and ultra-dispersed fluoropolymer particles into the melt of fibre-forming polymer in the process of yarn formation, or by treating the surface of polymeric material with gaseous fluorine. The use of these methods should ensure the formation of a nanoscale fluoropolymer film on the surface of the fibrous material, firmly bound to the macromolecules forming it, the thickness of which ensures the strength and durability of the formed coating.

A fundamentally new method of obtaining composite olifa by copolymerisation of dichloroisobutylenes contained in the cube waste of 3-chloro 2-methylpropene production with a-methylstyrene has been developed. On the basis of this work, a new anticorrosive composition for the protection of petrochemical equipment, possessing waterproofing and antiseptic properties, based, as mentioned above, on 3-chloro-2-methylpropene and bitumen, was developed.

In many highly aggressive and high-temperature environments in chemical plants, products such as, for example, chemical gaskets, flange seals, diaphragms, shut-off valve diaphragms and other products made of rubber or plastics are inoperable, and companies are forced to use, for example, rubber gaskets in fluoroplastic covers as sealing materials. They also fail quickly.

The use of highly chemical-resistant fluoroplastic as gasket and sealing materials will also not give positive results due to its fluidity under load. In this connection new highly effective composite corrosion- and heat-resistant gasket-sealing and protective elastic polymeric materials of BS-45 (VS-41, BS-43) type, workable in the temperature range of 60-350 °C and heat-resistant materials (B-800, B-850 type), workable in the temperature range from - 60 to 375 °C were developed to replace the existing obsolete analogues - rubbers, plastics, paronite, ebonite, asborezinovyh materials and others. The task of creation of such curing elastic rubber-plastic composite materials was solved due to the application of a fundamentally new type of polyolefin rubber first developed by us. Quickly vulcanising chlorinated corrosion- and ozone-resistant ethylene-propylene copolymer combines well (unlike the initial elastomer) with all types of highly unsaturated rubbers. In these compositions new thermo-corrosion-resistant chlorinated polyolefins, a number of synthetic resins, catalysts, special hardeners, fillers, softeners and other components are used.

Chlorinated polyethylene is manufactured in different countries and by several companies. The USA produces CPE of "Plaskon" and "Tyrin" trade marks, England - "Haloflex", Japan - "Elaslen", FRG produces homogeneous mixture of CPE and PVC of "Hostalit Z" trade mark.

Non-combustible floors are manufactured using CPE or a mixture of CPE and PVC with antimony oxide added as a flame retardant. CPE ensures compatibility between PVC and antimony oxide.

Currently, the main area of application of chloropolyethylene is its use as an additive to PVC to improve various properties. Of particular importance is the use of CPE as a high molecular weight plasticiser to improve the impact strength and elasticity of PVC. Compared to low molecular weight plasticisers, CPE does not migrate in PVC mixtures, it is not volatile, so that the plasticised composition has a significantly longer service life. Depending on the ratio of PVC and CPE, the impact strength, frost resistance, elasticity of the composition change. At high content of CPE the tensile strength decreases. The elasticity of the composition with increasing content of CPE increases up to a certain composition of the mixture, then this increase is insignificant. The composition is also characterised by anisatropy of mechanical properties.

CPE produces flexible semi-rigid non-combustible foams that are superior to polyurethane and polystyrene foams in terms of impact and abrasion resistance.

Laminated materials characterised by high flexibility, chemical and weather resistance are produced by combining a film or sheet of CPE material with fabric or foam. The resulting sheet laminated material is used, for example, for lining of tanks for storage and transport of liquids, particularly petroleum products.

An important area of application of CPE is obtaining fire-resistant sheathing for cables. The combination of dielectric properties with abrasion resistance, ozone, heat, oil and fire resistance make this electrical insulating composition suitable for use in various conditions.

The introduction of even small amounts of CPE into ABS plastics can dramatically reduce thermo-oxidative degradation and improve the mechanical properties of the plastics.

Modification of CPE, obtained from low density polyethylene, with copolymers of vinyl chloride and butyl acrylate, 2-ethylhexyl acrylate and polyvinyl chloride makes it possible to increase the strength of the material without reduction of elasticity and deterioration of frost resistance.

Grafted copolymers of CPE with styrene and acrylonitrile are widely spread abroad. These polymers, called ACS, are characterised by increased weather resistance, impact resistance, chemical resistance, heat resistance, very low flammability and good antistatic properties compared to ABS plastic, which competes with it in various fields of application.

It is established that the use of ammonium chloride compounds, have the ability to reduce the viscosity of paraffin oil six and more times. They can be recommended as an active base for creation of a reagent facilitating oil transportation.

As can be seen from the above, polymer composite materials containing halogen derivatives are characterised by high plasticity, fire resistance, frost resistance, which makes them indispensable in a number of industries.

CHAPTER 2

OBJECTS AND METHODS OF EXPERIMENTAL
RESEARCH

This chapter characterises the materials used in the study, specimen preparation techniques and describes the test methods used in the work presented.

2.1. Characterisation of source materials

High-pressure polyethylene.

Density g/cm3- 0.9180 - 0.9220

PTR g/10min - 0.7 - 1.50

Voltage coefficient - 1.30 - 1.38

Recommended application area - for general purpose films and film products

Low-pressure polyethylene

Density g/cm^3 - 0.9585 - 0.9620

PTR g/10min - 7

Voltage coefficient - 1.23 - 1.33

Processing method - injection moulding

Recommended application - for the manufacture of containers, baskets, crates, etc.

Recycled polyethylene and low molecular weight polyethylene

Hydrochloric acid is a solution of hydrogen chloride in water (27.5 - 38% by mass), boiling point t_M n - 108.6° C

Potassium permanganate - KMnO4, dark purple crystal, decomposition temperature trasl - 240° C.

Toluene - melting point $\wedge_{лав}$ - 95° C, boiling point $t^\wedge$ - 110.6° C, relative density at 20° C - 0.8669 g/cm^3 .

Diethanolamine - (NOSNZSN^INmelting point t_{ra} 27.8° C, boiling point 270° C, density at 20° C 1.0919 g/cm^3 . It is used in production as plasticisers, surfactants, dispersants for paints, corrosion inhibitors, absorber of acid gases from industrial gas mixtures.

Triethanolamine - (HOCH2CH?)3N melting point t_{ra} 21,2° C, boiling point 360° C. It is used in the production of soaps, detergents, as surfactants, corrosion inhibitors, absorber of acidic gases from industrial gas mixtures.

2.2. METHODOLOGY FOR PREPARATION OF REAGENTS AND SAMPLES
2.2.1. PURIFICATION OF THE SUBSTANCES USED

Distillation (at atmospheric pressure) was carried out to purify the solvent. An apparatus for distillation at atmospheric pressure is used, consisting of a Wurtz flask (with a tube soldered to the neck of the flask to remove the vapours of the boiling liquid to the refrigerator), a separate flask fitted with a Wurtz nozzle, a thermometer, a descending refrigerator, a flonge and a receiver.

Pieces of adobe porcelain, bricks and also glass capillaries patched on one side, which are placed in the flask with the open end towards the bottom, are put into the flask to influence the distilled liquid for even boiling.

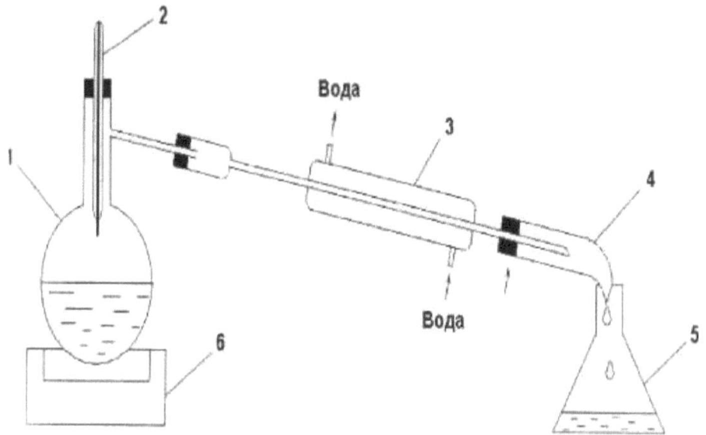

Fig. 2.1. 1-Würtz flask; 2-thermometer; 3-refrigerator. 4-column; 5-receiver; 6-electric cooker.

The distillation flask is heated on a closed electric cooker, a gas burner through an asbestos mesh, a flask heater, or on a water bath.

2.2.2.2.PURIFICATION OF LOW MOLECULAR WEIGHT POLYETHYLENE

Рис.2.2

For purification of low molecular weight polyethylene, a solution of polyethylene in cyclohexane was taken, which in such a form comes from the Shurtan gas chemical complex.For this purpose, a device called "Sakslit"   (Fig.2.2)   is filled with a solution of polyethylene in cyclohexane.

of low molecular weight polyethylene in cyclohexane and a reflux condenser is attached to the apparatus.

This dish with the solution is put on a flask filled with acetone. The heating temperature is chosen so that only acetone boils (t^ - 56° C). In the course of evaporation and reverse cooling acetone will wash out low molecular weight polyethylene and it will be collected in a special compartment of the saxlite. In this way, the clean low molecular weight polyethylene can be carefully removed from the instrument at the end of purification.

2.2.3. DETERMINATION OF SOLUBILITY

The determination of solubility of polymers is of great practical importance in their processing, as well as in the study of the features of their structure.Polymer dissolution occurs when the total energy of interaction between polymer molecules and solvent molecules exceeds the interaction energy between polymer molecules and between solvent molecules.The process of polymer dissolution has a characteristic feature: dissolution is usually preceded by swelling, accompanied by an increase in the volume of the polymer.

Quantitative determination of solubility.

In clean dried test tubes place 10 mg of dry crushed polymer and pour 2 ml of solvent from a microtube or graduated pipette. The tubes are corked and kept for 24-48 h at room temperature, shaking their contents periodically. The polymer is considered soluble if a homogeneous solution is formed. If the polymer does not dissolve at room temperature, the mixture is carefully heated to the boiling point of the solvent.

Comparison of solvent solubility of solvents.

Sometimes it is necessary not only to know in which solvents a given polymer is soluble, but also to estimate their solubility. The solubility of a solvent with respect to a given polymer can be measured by the dilution value, i.e. the amount of precipitant (non-solvent) added to the solution until the solution appears non-vanishingly turbid. Solutions prepared for the qualitative determination of solubility can be used for the determination.

For this purpose, the precipitant is added slowly drop by drop to the test tube with the solution from a microtube while stirring. The addition of the precipitant (titration) is stopped when a non-extinguishable turbidity appears in the solution. The amount of precipitant (in ml or in mol/mol of solvent) consumed for the titration serves as a measure of the solubility of the polymer. The solvents are compared on the basis of their values. More precipitant must be added to the polymer solution that contains the better solvent

2.3. CHARACTERISATION OF RESEARCH METHODS
2.3.1. PRODUCTION OF CHLORINATED POLYETHYLENE BY CHLORINATION IN SOLUTION

In order to carry out the chlorination reaction, chlorine must first be synthesised.Chlorine is synthesised by the action of concentrated hydrochloric acid on potassium permanganate.The reaction proceeds according to the following scheme:

$$2KMnO_4 + 16HCl = 2KCl + 2MnCl_2 + 5Cl_2 + 8H_2O$$

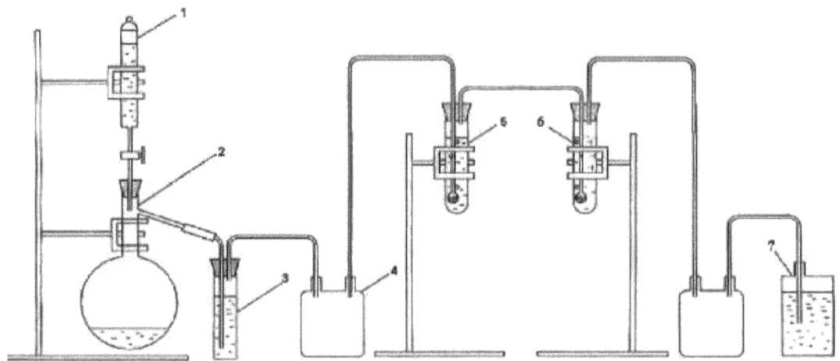

Fig.2.3 Installation for chlorination of polyethylene: 1 - separating funnel; 2 - round bottom flask; 3 - trap; 4 - safety device; 5,6 - reaction flask; 7 - neutraliser.

For the synthesis is used installation Fig.2.3.in a round bottom flask (2) placed the desired amount of KMnO4, which is affected by drops of hydrochloric acid coming through a separating funnel (1) with a ground stopper.

The chlorine supplied to the synthesis is passed through a test tube (3) half filled with sulphuric acid. The tube plays the role of a dryer, retains water vapours.The dried chlorine passes through the Tishchenko's vial - a kind of fuse (4), preventing the sulphuric acid from getting into the reactor (5), which contains the polyethylene solution.Then the remaining chlorine after the reaction is fed into the next reactor (6) for intensive chlorination. The remaining chlorine is passed through another fuse, at the very end of the installation the tube is immersed in a beaker or flask with alkaline solution (7), to neutralise the unreacted chlorine.

The reaction is carried out in a fume cupboard.

Operating principle of the chlorinator

In order to establish optimal conditions for chlorination, the synthesis was carried out under different conditions. For example, in a round bottom flask equipped with a stirrer. In this case, free chlorine in the system is obtained due to the interaction of potassium permanganate with concentrated hydrochloric acid, and then the experiments were transferred to a vertical unit with a barbater, to which was later added a stirrer.Equipment: chlorinator with a barbater pipe for gas removal, reflux condenser, UV lamp, oil bath.

3% polyethylene solution is loaded into the chlorinator and heated on an oil bath to 70-80 °C; at this temperature chlorine starts to flow through at a rate of 0.5 l/min. The chlorinator is irradiated with UV lamp light. Chlorination is carried out for 6 hours.

The prepared polymer is separated from the liquid by decantation, washed with alcohol and dried in a thermostat.

2.3.2. PRODUCTION OF CHLORINATED POLYETHYLENE BY SLURRY CHLORINATION METHOD

This method became widespread after the emergence of high-density polyethylene. Chlorination in suspension can be carried out in various media: in water, acetic acid, cold carbon tetrachloride.

Chlorination in suspension was carried out according to the following method. Polyethylene was loaded into a chlorinator, a surfactant and water were added and saturated with hydrochloric acid. The reaction mass was also irradiated with ultraviolet rays. Upon completion of the reaction, the chlorine gas residue was purged with nitrogen current, filtered and the filtrate was washed with alcohol, and then also dried in the thermostat at 50-60° C.

2.4. DETERMINATION OF CHLORINE IN POLYOLEFINS BY THE CARIUS METHOD

This method is based on the determination of halogens and sulphur in organic compounds. The analysed substance is decomposed by fuming HNO3B in a sealed tube at 200-300° C in the presence of AgNO3 (in the determination of halogens) or BaCl2 (in the determination of sulphur) after which the tube is opened and the elements are determined gravimetrically (as Ag or $BaSO_4$ halides, respectively).

2.5. DETERMINING THE MELTING POINT OF THE POLYMER

The polymer is crushed and 1 g is placed on a metal plate heated from below with a burner or electric cooker. (The plate has a hole in the side, into which a thermometer is inserted.) The plate is gradually heated, and the polymer is constantly stirred with a glass stick. The temperature at which the polymer begins to stick together is taken as the softening point.

The conditional melting point of the polymer (since the polymer is polydisperse and does not melt at the point) can be determined by the usual method used to determine the melting point of low molecular weight substances, using a capillary. The finely ground polymer is placed in a capillary (40-50 mm long, sealed at one end), which is fastened to the thermometer with a rubber ring. (The polymer in the capillary should be at the level of the mercury ball of the thermometer.) The thermometer with the capillary is placed on the stopper in a dry test tube so that it does not touch the walls, the lower end of the thermometer should be 0.5-1 cm above the bottom of the test tube. The test tube with the thermometer and capillary are placed in a beaker filled with a high-boiling liquid (silicone oil, glycerine, concentrated sulphuric acid, etc.). The beaker is gradually heated (heating rate $\sim 1^0$ C per minute) and the change of the polymer in the capillary is observed. The temperature at which the polymer completely melts in the capillary is taken as the conditional melting point of the polymer.

2.6. PHYSICOCHEMICAL METHODS OF ANALYSIS OF SYNTHESISED OLIGOMERS

IR spectra of the initial and synthesised compounds were taken on a spectrophotometer (FTIR) model NEXUS 470. The samples were taken in the form of films.

Thermodegradation of the synthesised compounds was investigated by differential-thermogravimetric analysis on the derivatograph system Paulik F, Paulik I, Erdey L, based on the change of thermal effects during heating of compounds in the temperature range of 20-800° C.

2.7. METHOD FOR DETERMINING RESISTANCE TO
CHEMICAL MEDIA

Determination of mass change of polymer samples, diffusion, sorption and permeability coefficients

For comparative tests of different polymers by these methods it is necessary to use samples of the same shape, the same dimensions, manufactured by the same technology, and to test them under the same regimes.

After weighing each specimen, they are placed in a vessel with a chemical reagent heated to the test temperature. The samples are placed in the vessel so that they are completely immersed in the chemical reagent. The duration of the test is selected according to GOST 12020-72.

The chemical reagent is stirred with a stirrer at least once a day during the test.

After the test is completed, the specimens are rinsed with a non-corrosive liquid.

The change in mass of the specimen after each test period (DM) in per cent weight gain or loss shall be calculated according to the formula:

$$\Delta M = \frac{(M_1 - M) \cdot 100}{M} \qquad (2.1)$$

Where:

M - mass of the test sample before its first immersion in the chemical reagent, g;

M1 - mass of the test specimen after soaking it in the chemical reagent, g;

The arithmetic mean of at least five determinations shall be taken as the result of each test. According to the obtained results, the graphical dependence is plotted $\Delta M = f(\tau)$.

From the graph determine the time for which there was an increase in the mass of the sample to the value $M_{Max}/2$, and calculate the diffusion coefficient of the chemical reagent in the polymer sample (D) in cm^2 /s by the formula:

$$Д = 0{,}0494 \left(\frac{\tau_0}{\delta^2}\right)^{-1} \qquad (2.2)$$

Where:

τ_0 — time during which there was an increase in the mass of the sample to $M_{Max}/2$,

M_{Max} - mass of the test sample at the established sorption equilibrium, g;

δ - sample thickness, cm.

Using the results calculate the sorption coefficient of the chemical reagent in the polymer sample (S) in g/cm^3 , according to the formula:

$$S = \frac{M_\rho}{V_{max}} \qquad (2.3)$$

Where:

V_{max} - volume of the test specimen after the end of the test, cm^3 ;

29

The mass of chemical reagent absorbed by the test sample (Mp) in g is calculated according to the formula:

$$M_\rho = M_{max} - M \qquad (2.4)$$

Where:

M_{max} - mass of the test sample at the established sorption equilibrium, g;

M - mass of the test sample before its first immersion in the chemical reagent, g.

Permeability coefficient of chemical reagent through polymer samples (P) in g cm/cm^2 is calculated according to the formula:

$$P = Д \cdot S \qquad (2.5)$$

Where:

D - diffusion coefficient, cm /cm;^{2}

S - sorption coefficient, g/cm^3 .

Changes in the appearance of samples are determined by visual comparison with a sample that has not been tested. Changes in colour, gloss, cracks, bubbles are determined.

It is recommended that the visual assessment of change in appearance be labelled as follows: O - no change, F-insignificant change, M - moderate measurement, L-significant change.

Preliminary assessment of polymer resistance to
chemical reagent exposure
by changes in mechanical performance.

Change of mechanical index (increase or decrease of it in comparison with the initial value) after the samples stay in the reagent (ΔG) in per cent is calculated by the formula:

$$\Delta G = \frac{(G_1 - G) \cdot 100}{G} \qquad (2.6)$$

Where:

G - arithmetic mean value of the determined mechanical index in the initial state before immersion in the reagent;

G1 - arithmetic mean value of the determined mechanical index after soaking the samples in the reagent.

The results of the calculations are rounded to whole numbers.

Preliminary assessment of the resistance of plastics to the action of chemical reagent is carried out according to the change of mechanical parameters of polymers in accordance with GOST 12020-72.

2.8. DETERMINATION OF RELATIVE ELONGATION AT
BREAK

Relative elongation at break is the change in the design length (A1$_{op}$) of the specimen during the tensile test at the moment of rupture, related to the initial design length (l_o)

of the specimen. To determine these values, the test shall be carried out on a machine which, when the specimen is stretched, shall be capable of measuring the load with an error of not more than 1 per cent of the measured value and a constant rate of extension of the clamps within the limits of the relevant standards.

In the research work the sample of type 5 according to GOST 11262-80 with reduced dimensions was chosen.

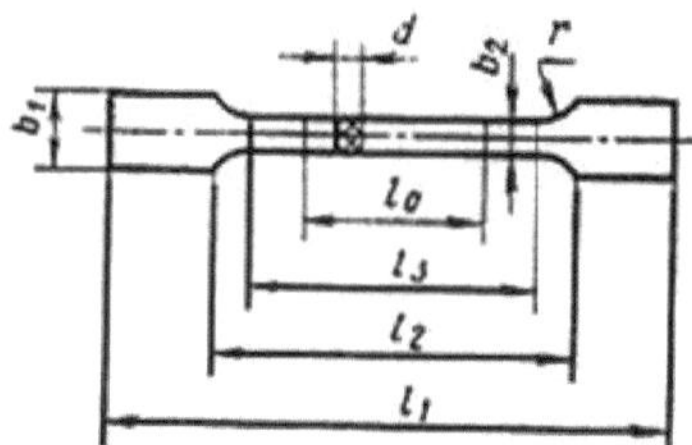

Fig.2.4 View of the sample according to GOST

When measuring relative elongation by the change in the distance between clamps, the measured relative elongation should be more than 10 per cent.

Before testing, the specimens shall be labelled in accordance with GOST. The marks shall not impair the quality of the specimens or cause rupture of the specimens in the places of marks. The arithmetic mean of at least five determinations calculated to the third significant digit shall be taken as the test result.

The relative elongation at break (ε_{pp}) is calculated using the following formula:

$$\varepsilon_{pp} = \frac{\Delta l_{op}}{l_0} \cdot 100; \qquad (2.7)$$

Δl_{op} - change in the design length of the specimen at the moment of rupture, mm;

l_0 - initial design length of the specimen, mm.

The arithmetic mean of at least five determinations rounded to two significant figures shall be taken as the test result.

CHAPTER 3

EXPERIMENTAL PART
3.1. STUDY OF POLYETHYLENE CHLORINATION

The industry of synthesis and processing of plastics, being a major supplier of polymer products to many industries, agriculture, services, trade, science, defence, engineering, has a significant impact on the scale, direction and efficiency of their development. This is due to the fact that every year the volume of growth in the needs of various industries, requires more and more polymeric materials to solve a variety of problems. One of the directions of obtaining new polymeric materials is their chemical modification. As it is known, one of the ways of modification of polyolefins is the use of chlorination. Polymer composite materials containing halogen derivatives are characterised by high plasticity, fire resistance, frost resistance at availability, which makes them indispensable in a number of industries. It is known that today chlorinated high and low density polyethylene, chlorinated polypropylene and chlorinated polyvinyl chloride are used in the production of plastics and elastomers, skin-like rigid and brittle materials.

The aim of this work is to investigate the chlorination process and to study the chemical transformations during the modification of chlorinated polyethylene.

By chlorination of polyolefins it is possible to obtain products that differ significantly in properties from the original polyolefins. Chlorinated polymers based on all industrial polyolefins - low and high density polyethylenes, isotactic and atactic polyethylene - have been described in the literature. However, the main industrial applications are modified polymers based on polyethylenes (PE) of different densities. Chloropolyethylene (CPE), can contain from 20 to 72% chlorine. The chlorination process under the action of molecular chlorine proceeds as a chain radical reaction.

The process is initiated under the action of radiation of visible or ultraviolet part of the spectrum, ionising radiation, substances decomposing into free radicals: organic peroxides or hydroperoxides (usually benzoyl peroxide), diazo compounds (azobisisobutyronitrile), organometallic compounds, iodine, metal chlorides (AlCl3, FeCl3, etc.).

$$\left[CH_2{-}CH_2\right]_n + nCl_2 \longrightarrow \left[CH_2{-}CH{-}CH_2{-}CH{-}CH_2{-}CH\right]_n$$
$$\qquad\qquad\qquad\qquad\qquad\qquad\quad |\qquad\qquad|\qquad\qquad|$$
$$\qquad\qquad\qquad\qquad\qquad\qquad\ Cl\qquad\ Cl\qquad\ Cl$$

Chlorination was carried out in a continuous process. Chlorination
The continuous method produces a more homogeneous product.

Chlorination was carried out in chlorinators equipped with mechanical stirrers.
Only at halogenation in solvent the polyolefin is not subjected to high requirements in terms of grinding, and it is possible to obtain a product with chlorine content of more than 70%. The disadvantage of this method is the difficulties associated with the separation of the product from dilute (5-7%) solutions and removal of traces of solvent

from it. In spite of the above, a more efficient and economical method is the process of chlorination in organic solvents.

Chlorination of LDPE, LDPE, recycled PE and low molecular weight PE was carried out in organic solvent. At room temperature PE is insoluble in common solvents. At elevated temperatures, depending on the degree of crystallinity, it dissolves in chlorinated aliphatic and aromatic hydrocarbons. Toluene was used as a solvent.

The temperature at the beginning of chlorination was 50-60^0 °C and at the end of chlorination 100-110 °C. Further increase of temperature led to worse results. Also increasing the duration of chlorination over 6 hours did not give positive results. The results of the conducted studies on qualitative determination of chlorine showed the presence of bound chlorine in the system. Chlorine content in the obtained samples was determined by the Carius method.

3.2. STUDY OF THE INTERACTION OF XPE WITH DIETHANOLAMINE AND TRIETHANOLAMINE

Chlorinated polyethylene with different chlorine content is quite well soluble in some lacquer solvents, thanks to which it is possible to achieve the viscosity of the composition required for use in the paint and varnish industry. However, it should be noted that coatings based on CPE with chlorine content up to 40% have become much less widespread than coatings based on chlorosulphated polyethylene. This is due to non-standardity of CPE batches, longer duration and higher curing temperature. Aliphatic and aromatic di- and polyamines are used as hardeners for CPE. The resulting films have a fairly high density and elasticity, but are characterised by high stickiness and slow dust drying. As a rule, hot-drying coatings are obtained from CPE. CPE-based coatings have high corrosion resistance in various aggressive environments.

Studying the interaction of CPE with diethanolamine and triethanolamine, it was concluded that the main direction of the reaction is dehydrochlorination with the formation of double bonds in polymer chains, amination and intermolecular imination.

As the chlorine content in the polymer increases, its tendency to dehydrochlorination reaction with the formation of C=C bonds increases. This is manifested in the deepening of the colour of the formed product. The proportion of nitrogen attaching to CPE is small.

Later it was found that amination of CPE occurs as a result of substitution of chlorine atoms at tertiary carbon atoms, since the content of attached nitrogen corresponds to the number of branching in the polymer chains. According to the IR spectroscopy data, the most significant changes in the structure of CPE are observed during the amination of diethanolamine and triethanolamine.

3.3. STUDY AND ANALYSIS OF SYNTHESISED OLIGOMERS BY IR SPECTROSCOPY

The method of infrared (IR) spectroscopy is currently one of the most widespread methods for identification of polymers and polymer composite materials and their structural analysis. The wide spread of this method in comparison with other modern physicochemical and chemical methods is due to the availability and reliability of modern serial IR spectrometers, high speed of analysis, high level of experimental technique of spectral study of polymer systems. In addition, the polymer does not undergo chemical changes during the process of analysis. The method of IR spectroscopy is based on the interaction of a substance with IR radiation in the wavelength region X=2.5-25 μm (4000-400 cm^{-1}). This region is called the mid-IR region. The region 400-10 cm^{-1} belongs to the far IR region, and the region 12500-4000 cm^{-1} - to the near IR region.

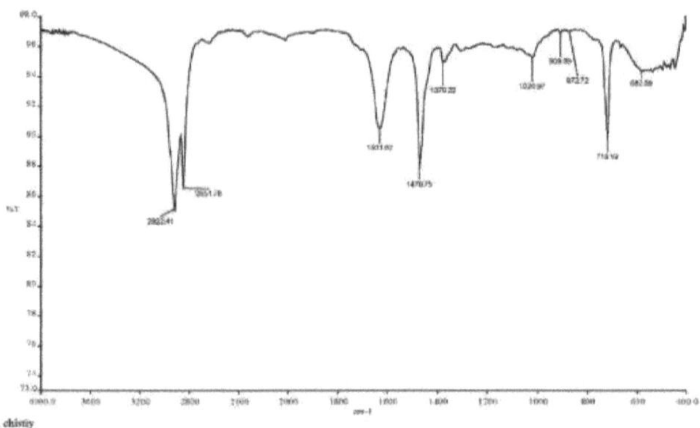

Fig.3.1 IR spectrum of LDPE

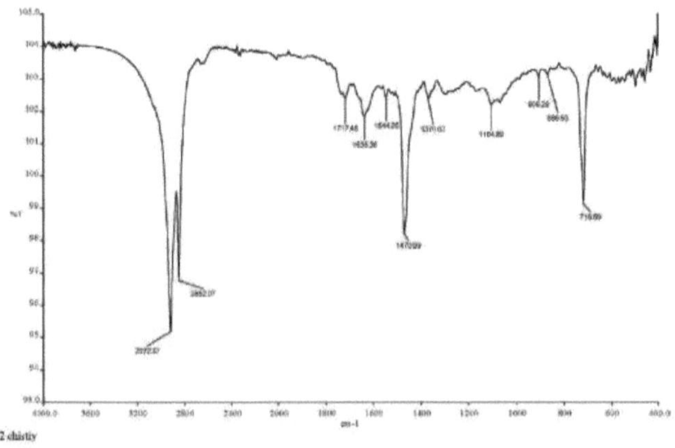

Fig.3.2 IR spectrum of PENDPE

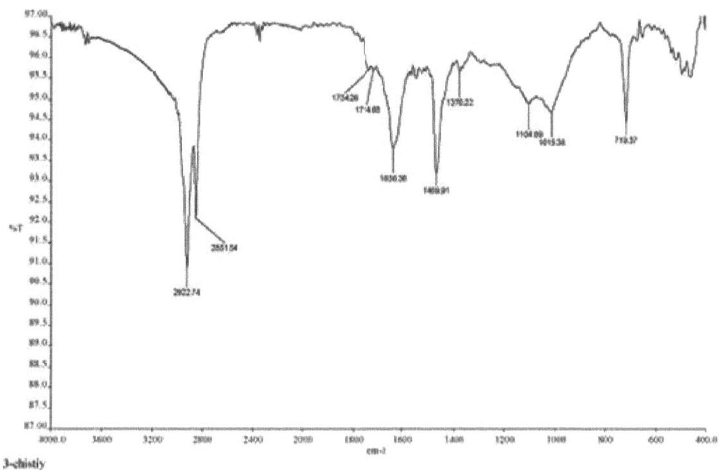

Fig. 3.3 IR spectrum of secondary PE

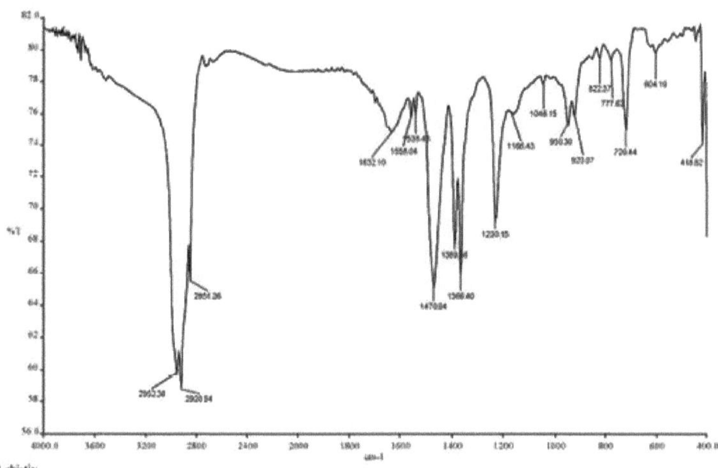

Fig. 3.4.IR spectrum of low molecular weight PE

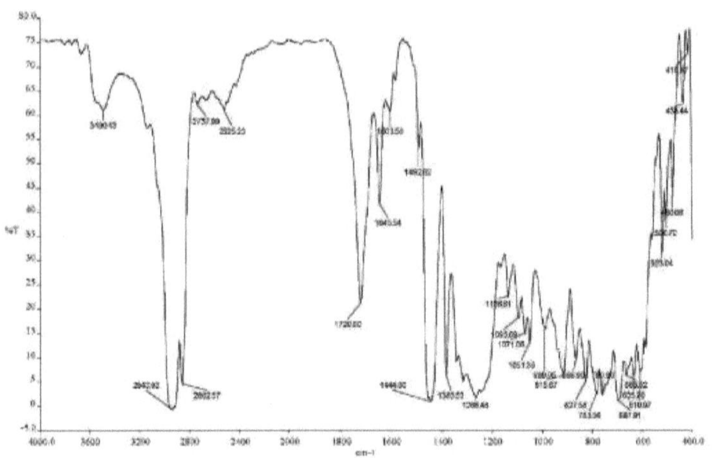

Fig.3.5 IR spectrum of chlorinated LDPE

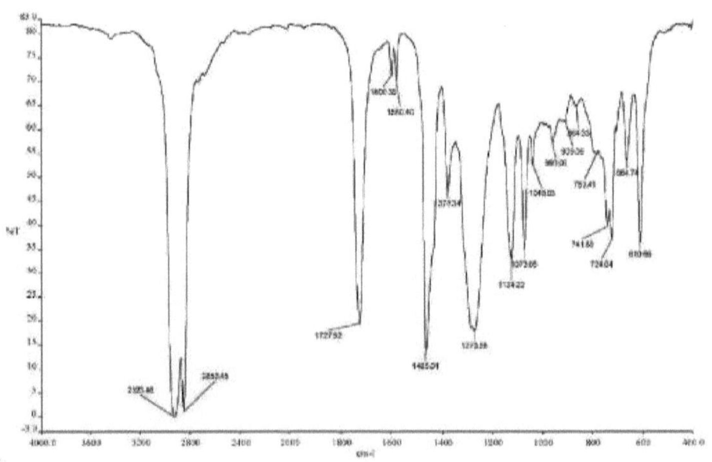

Fig.3.6 IR spectrum of chlorinated HDPE

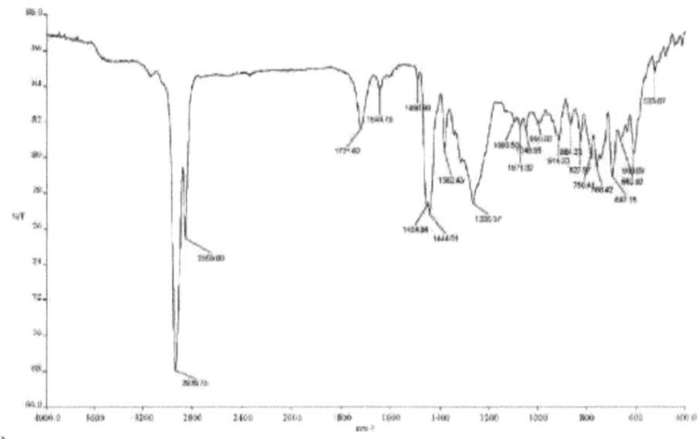

Figure 3.7. IR spectrum of chlorinated secondary PE

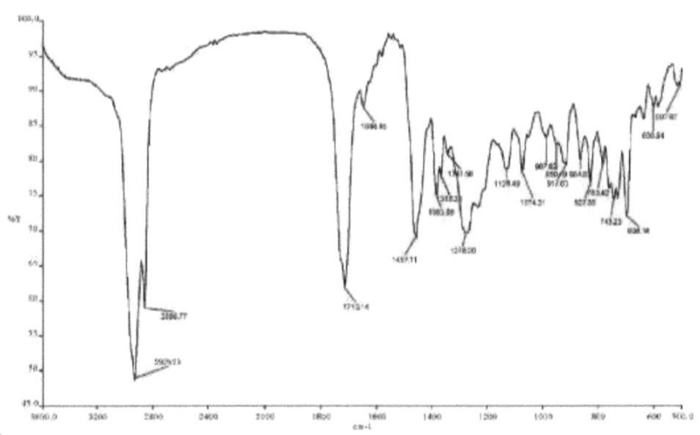

Fig.3.8: IR spectra of chlorinated low molecular weight PE

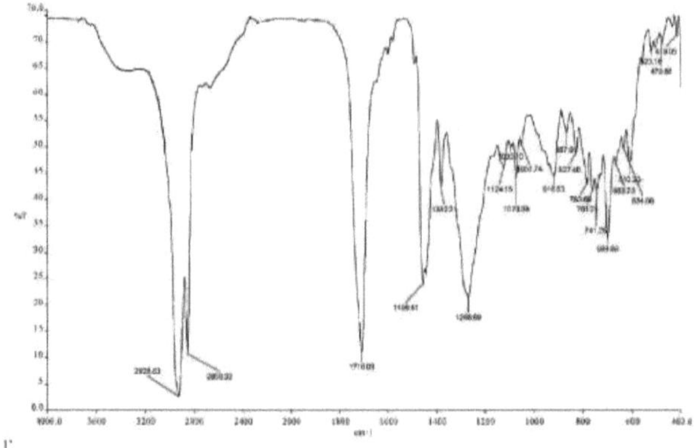

Fig.3.9. Infrared spectra of chlorinated LDPE modified by

диэтаноламином

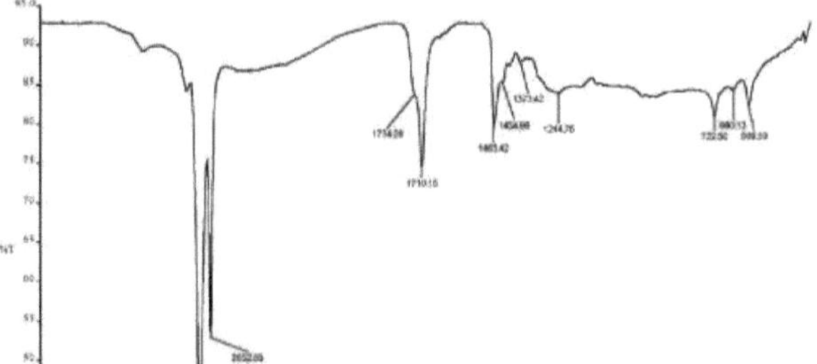

Fig.3.10. IR spectrum of chlorinated HDPE modified with diethanolamine

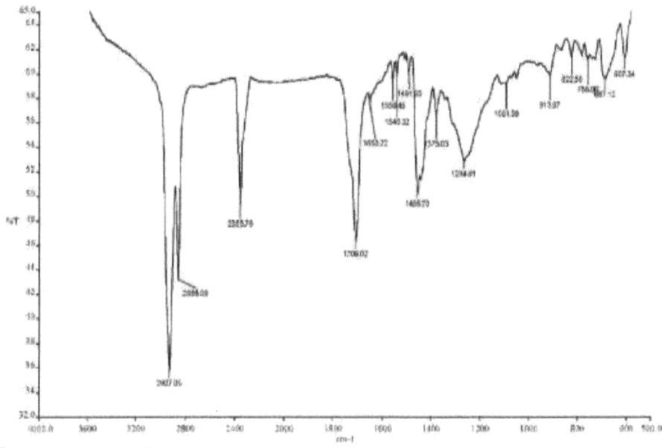

Fig.3.11. IR spectrum of chlorinated secondary PE modified with diethanolamine

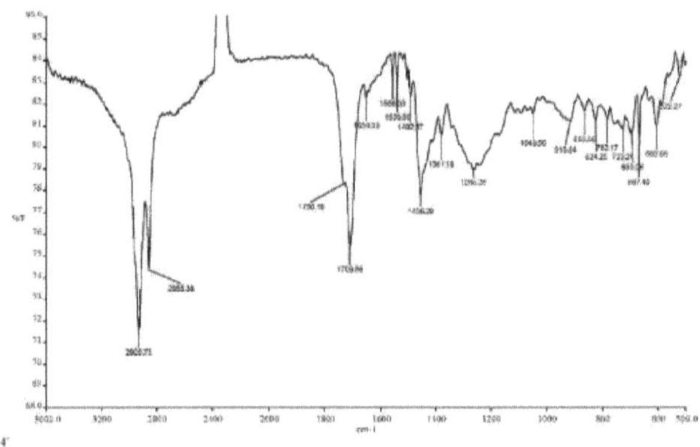

Fig.3.12. IR spectrum of chlorinated low molecular weight PE modified with
diethanolamine

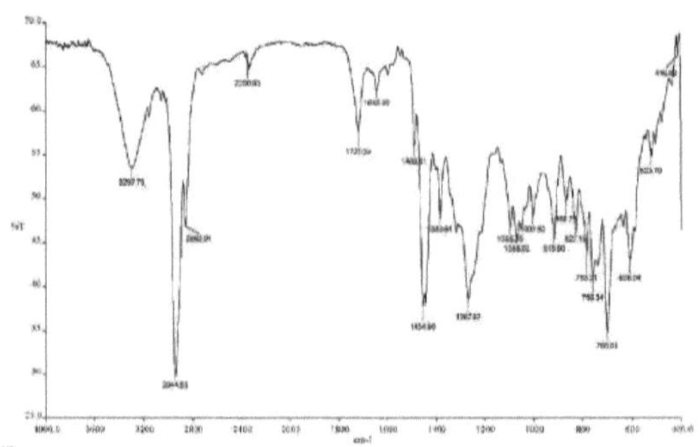

Fig.3.13. IR spectrum of chlorinated LDPE modified by

triethanolamine

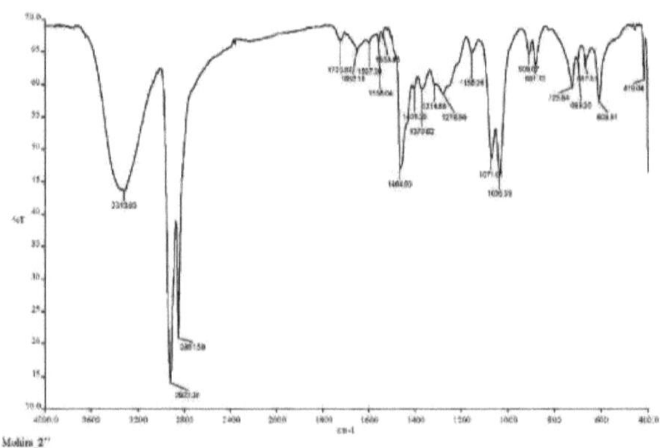

Fig.3.14. Infrared spectrum of chlorinated HDPE modified by
triethanolamine

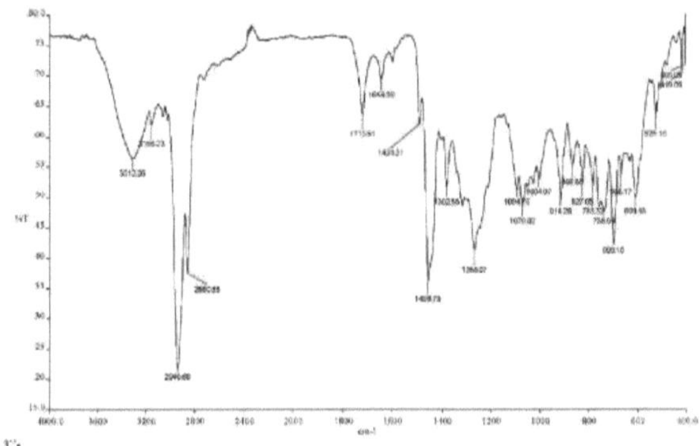

Fig.3.15. IR spectrum of chlorinated secondary PE modified with triethanolamine

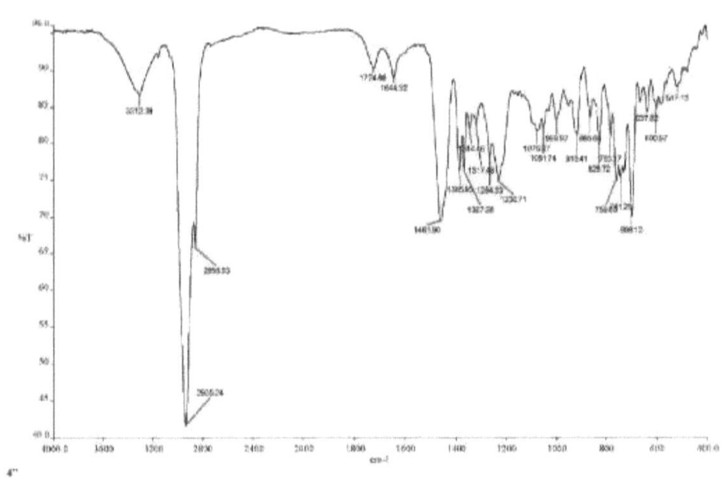

Fig.3.16. IR spectrum of chlorinated low molecular weight PE modified with
triethanolamine

IR absorption spectra of organic ligands

Figures 3.1. - 3.16. IR absorption spectra of high pressure polyethylene (HPPE) molecule, chlorinated HPPE, chlorinated HPPE modified with diethanolamine, chlorinated HPPE with triethanolamine, low pressure polyethylene (LPE), chlorinated LPE, chlorinated LPE modified with diethanolamine, chlorinated LDPE with

41

triethanolamine, secondary polyethylene (SPE), chlorinated SPE, chlorinated SPE modified with diethanolamine, chlorinated SPE with triethanolamine, low molecular weight polyethylene (LMPE), chlorinated LMPE, chlorinated LMPE modified with diethanolamine and chlorinated LMPE with triethanolamine. Table 3.1. shows the values of characteristic frequencies (see^{-1}) of the above compounds.

The IR absorption spectrum of high pressure polyethylene(HDPE) molecule Fig.-3.1.is characterised by bands at 2922 - v_{as} (CH$_2$), 2852 - vs(CH$_2$), 1632- v(C=C), 1471 - 6 (CH2), 1376, 1021 - v(C-C), 909 - n(CH), 873 - v(CC), 719 and 582 cm 1.

The IR absorption spectrum of the low-pressure polyethylene (LDPE)molecule Fig.-3.2. shows frequencies at 2922 - Vas(CH2), 2852 - vs(CH2), 1717, 1636 - v(C=C), 1544, 1471 - 6(CH$_2$), 1371,1105, 906 - n(CH$_2$),870 - v(CC) and 719 cm^{-1} .

The IR absorption spectrum of the secondary polyethylene (SPE) molecule of Fig.-3.3. shows frequencies at 2923 - vas(CH2), 2852 - vs(CH2), 1734, 1715, 1636 - v(C=C), 1470 - 6(CH2), 1376 -(CH2), 1105, 1015 and 719 cm^{-1} .

The IR absorption spectrum of low molecular weight polyethylene (LMWPE) Fig.-3.4.is characterised by bands at 2952 - vas(CH2), 2921 - vas(CH2), 2851 - vs(CH2), 1632 - v(C=C), 1558, 1538, 1471 - 6(CH2), 1390, 1366, 1230, 1166, 1046, 1046, 950, 923 - n(CH$_2$), 822 - v(CC), 778, 720, 604 and 419 - 6(CH$_2$) cm^{-1} .

In the IR absorption spectrum of chlorinated LDPE Fig.-3.5.frequencies (see^{-1}) were detected at 3490, 2944 - vas(CH2), 2863 - vs(CH2), 2738, 2525, 1721, 1621, 1646, 1604, 1493 - 5(CH$_2$), 1445 - 5(CH$_2$), 1384, 1266, 1137, 1096, 1071, 1051 - (C-C), 989, 916 - n(CH$_2$), 867, 828 - v(CC), 784, 761 - (C-Cl), 698, 666, 636, 611, 507, 480, 438 and 419 - 5(CH).$_2$

The IR absorption spectrum of chlorinated HDPE molecule Fig.-3.6.has frequencies at 2924 - Vas(CH2), 2853 - 5(CH$_2$), 1728, 1600 - v(C=C), 1580 - v(C=C), 1465 - 5(CH$_2$), 1378 - (CH2), 1274, 1124, 1073, 1040, 960, 909 - n(CH2), 864 - v(CC), 780, 742 - v(C-Cl), 724 - (C-Cl), 664 and 611 capac 1.

The IR absorption spectrum of VPE molecule chlorinated Fig.-3.7.has frequencies at 2940 - vas(CH2), 2860 - vs(CH2), 1722, 1645 - v(C=C), 1491, 1454 - 5(CH$_2$), 1444, 1382 - (CH$_2$), 1265, 1096, 1071, 1049, 998, 914 - n(CH$_2$), 864 - v(CC), 828 - v(CC), 780 cm^{-1} .

In the IR absorption spectrum of chlorinated NMPEs Fig.-3.8.frequencies were detected (see^{-1}) at 2930 - vas(CH2), 2857 - vs(CH2), 1715, 1646 - v(C=C), 1457 - 5(CH$_2$), 1384 - (CH2), 1366 - (CH2), 1342 - (CH2), 1278, 1128, 1074, 988, 950, 950, 917 - n(CH$_2$), 864 - v(CC), 827 - v(CC), 783, 743, 696, 601 and 508.

IR spectrum of chlorinated LDPE molecule modified with diethanolamine Fig.-3.9. characterised by bands at 2929 - vas(CH2), 2856 - vs(CH$_2$), 1710 - v(NH), 1456 - 5(CH$_2$), 1382 - (CH2), 1269, 1124, 1094, 1073, 1052, 916 - n(CH2), 867, 827 - v(CC), 784, 761, 741 - v(C-Cl), 700 - (CCl), 666, 635, 610, 523, 480 and 419 - 6(CH$_2$) cm^{-1} .

IR spectra of the chlorinated HDPE molecule modified with diethanolamine Fig.-3.10.characterised by bands at 2924 - vas(CH2), 2853 - vs(CH$_2$), 1734, 1710, 1465 - 6(CH$_2$), 1435 - v(C=C), 1373 - (CH$_2$), 1245, 722 - (CCl), 660 and 610 cm^{-1} .

IR spectra of the chlorinated WPE molecule modified with diethanolamine Fig.-3.11.characterised by bands at 2927 - vas(CH2), 2856 - vs(CH$_2$), 2356, 1709, 1653 - 5(NH$_2$), 1556 - v(C=C), 1540, 1491 - v(C=C), 1456 - 5(CH$_2$), 1379 - (CH$_2$), 1264, 1091, 914 - n(CH2), 823 - v(CC), 758, 687 and 607 cm^{-1} .

IR spectrum of chlorinated NMPE modified with diethanolamine Fig.-3.12.characterised by bands at 2926 - v$_{as}$ (CH$_2$), 2855 - v$_s$ (CH$_2$), 1730, 1710, 1651 - 5(NH2), 1557 - v(C=C), 1540, 1493, 1456 - 3(CH$_2$), 1381, 1265, 1049, 916 - n(CH$_2$), 864 - v(CC), 824 - v(CC), 782, 725, 693, 667, 604 and 522 cm^{-1} .

In the IR absorption spectrum of chlorinated LDPE modified with triethanolamine Fig.-3.13.Frequencies were found at 3298 - v(NH), 2944 - vas(CH2), 2863 - vs(CH2), 2351, 1720, 1646 - v(C=C), 1490, 1455 - 3(CH$_2$), 1384, 1268, 1095, 1068, 1004, 967 - n(CH$_2$), 827 - v(CC), 783, 759 - v(C-Cl), 700 - (CCl), 606, 524 and 417 - 3(CH$_2$) cm 1.

In the IR absorption spectrum of HDPE chlorinated modified with triethanolamine Fig.-3.14. Frequencies at 3314 - (O-H), 2922 - vas(CH$_2$), 2852 - vs(CH$_2$), 1726, 1653 - 3(NH$_2$), 1597 - v(C=C), 1558 - v(C=C) were found, 1538, 1465 - 3(CH$_2$), 1401, 1371 - (CH), 1315 - (CH), 1279, 1155, 1071, 1035, 909 - n(CH$_2$), 881 - v(CC), 724, 699, 668, 610 and 419 cm 1.

In the IR absorption spectrum of WPE chlorinated modified with triethanolamine Fig.-3.15. We found frequencies at 3312 - v(NH2), 3156 - v(NH$_2$), 2941 - vas(CH$_2$), 2861 - vs(CH$_2$), 1719, 1646 - 3(NH$_2$), 1491 - v(C=C), 1456 - 3(CH$_2$), 1382 - (CH$_2$), 1268, 1094, 1070, 1004, 914 - n(CH$_2$), 867- v(CC), 827 - v(CC), 783, 738, 699, 668, 609, 525, 419 - 3(CH$_2$) and 405 cm 1.

In the IR absorption spectrum of chlorinated NMPE modified with triethanolamine Fig.-3.16. Frequencies were found at 3312 - v(NH), 2935 - vs(CH$_2$), 2857 - vs(CH$_2$), 1725, 1646 - 3(NH$_2$), 1462 - 3(CH$_2$), 1386, 1367, 1344, 1317, 1264, 1231, 1077, 1052, 1000, 916 - n(CH$_2$), 866 - v(CC), 829 - v(CC), 783, 760, 741, 698, 637, 601 and 517 see^{-1} .

Table 3.1.

Values of characteristic frequencies (cm^{-1}) in the IR absorption spectra of LDPE, LDPE, LDPE and NMPE pure, chlorinated, chlorinated and modified with diethanolamine and triethanolamine

Connection	Vas (CH)$_2$	v(CH)$_2$	v(C=C), v(CC)	v(C-Cl)	5(NH2), v(NH), Vs(NH)
cavity	2922	2852	1632		
Chlorinated LDPE	2944	2863	1646, 828	761	
Chlorinated LDPE with diethanolamine	2929	2856	827	741, 700	1710
Chlorinated LDPE with triethanolamine	2944	2863	1646	759	3298
pend	2922	2852	1636		
Chlorinated HDPE	2924	2853	1600	742	

Chlorinated HDPE with diethanolamine	2924	2853	1435	722	1710
Chlorinated HDPE with triethanolamine	2922	2852	1597	724	1653
inpe	2923	2852	1636		
Chlorinated WPE	2940	2860	1645	780	
Chlorinated WPE with diethanolamine	2927	2856	1556	758	1653
Chlorinated WPE with triethanolamine	2941	2861	1491, 867, 827	738	3312, 3156
nmpe	2952, 2921	2921	1632		
Chlorinated NMPE	2921	2857	864	743	
Chlorinated NMPE with diethanolamine	2930	2855	864, 824	725	1651
Chlorinated NMPE with triethanolamine	2926	2857	866, 829	741	3312, 1646

In the IR absorption spectrum of LDPE, LDPE, LDPE, WPE and NMPE, the frequencies at 2922-2944 cm^{-1} , correspond mainly to the valence v_{as} (CH2) and v_s (CH2) bond vibrations. While the value of 1651-1710 cm^{-1} correspond at the v(NH) vibrational frequencies. From 700-780 cm^{-1} vibrational frequencies correspond on (C-Cl). With the transition to the modification state, i.e. in polymer compounds polyethylene molecules with amines are connected through carbon and chlorine atoms.

3.4. STUDY OF SYNTHESISED OLIGOMERS
BY DIFFERENTIAL THERMAL ANALYSIS
(DTA)

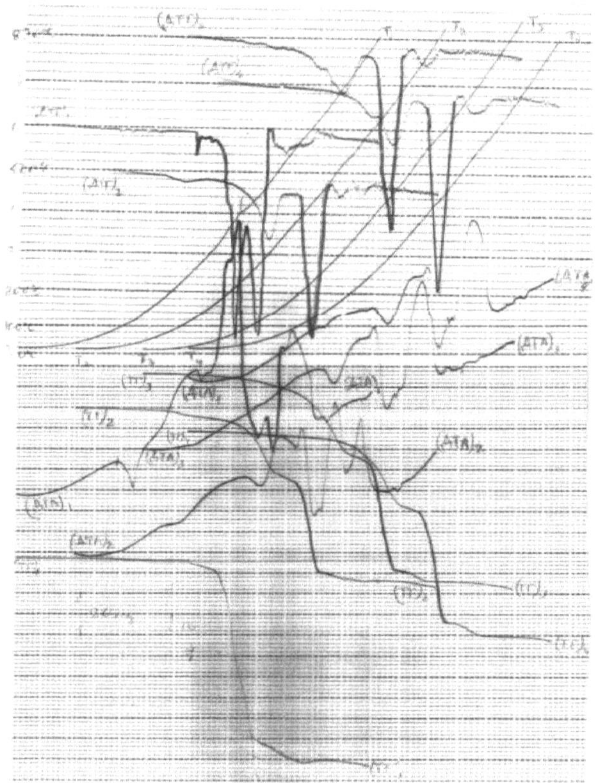

Figure 3.17: Derivatogram of PE, Chlorinated PE, CPE modified with diethanolamine and thiethanolamine.

The studies were carried out on a derivatograph of the Paulik-Herdey system of the MOM company (Hungary) in the temperature range 20-800° C. Using the derivatograph, mass loss (ML), mass change rate (MTR), thermal effects (DTA curve) and temperature changes (MTR curve) were simultaneously determined on one sample Fig.3.17.

Derivatograms of pure PE, chlorinated PE and chlorinated polyethylene modified with diethanolamine and triethanolamine were taken to compare the changes occurring in chlorinated polyethylene after addition of diethanolamine and triethanolamine.

Thermogravimetric curve (Fig.3.18.) was used to determine the thermal stability (heat resistance) of polymer samples.

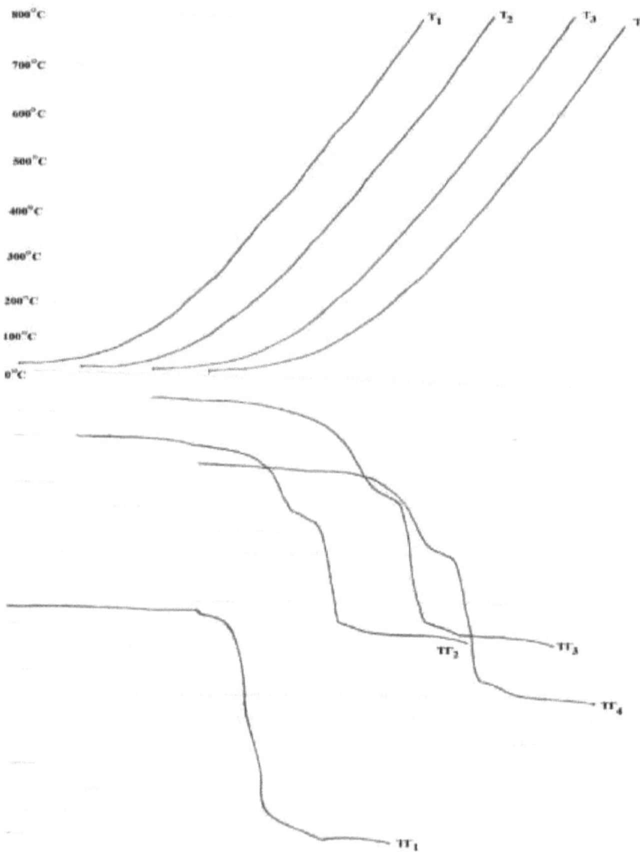

Fig.3.18.The curve of dependence of mass change on temperature (thermogravimetric curve) for the investigated samples.

The table shows the temperature of the beginning of polymer decomposition (T_n), the temperature at which the mass loss in % occurs (T_{10}, T_{20}, T_{50}) and the temperature at which the complete decomposition of the sample occurs, i.e. the final decomposition temperature (T_k):

Table 3.2.

Name of polymer sample		t_H ,°c	Tue,°C	T_{20} ,°C	t_{50} ,°c	Tk,° C
TG1	Low pressure polyethylene (unmodified)	283	365	395	410	680
tg₂	Chlorinated polyethylene	240	255	278	380	610
TG₃	Chlorinated polyethylene modified with diethanolamine	120	200	275	420	600
TG4	Chlorinated polyethylene	130	240	280	420	600

	modified with triethanolamine					

As can be seen from Table 3.2. the temperature of the onset of decomposition Tn for unmodified polyethylene is much higher (275° C) than for chlorinated polyethylene and modified chlorinated polyethylene samples (in the range 240-130° C). This implies that chlorinated polyethylene is less thermostable than pure polyethylene. Also modification of CPE with diethanolamine and triethanolamine did not give good results.

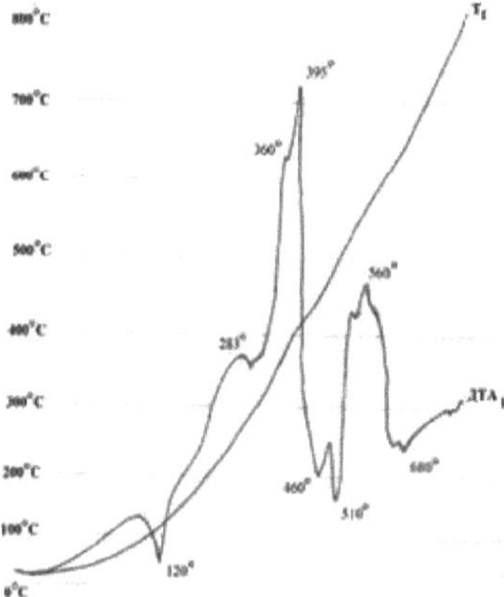

Fig.3.19. - DTA curve of polyethylene

As can be seen in the DTAl curve (Fig.3.19), exothermic effects at 283° C, 360° C, 395° C and 560° C and endothermic temperature changes at 120° C, 460° C, 510° C and 680° C are clearly detected. The exothermic effect at 120° C is related to the onset of softening temperature and the exothermic effects at 360° C, 395° C are related to the onset of thermo-oxidation of polyethylene. Apparently, in the initial stage of thermo-oxidation there is a breakdown of cross-links. The next exothermic effect at 560° C corresponds to the second stage of thermooxidation and the beginning of the polymer depolymerisation process.

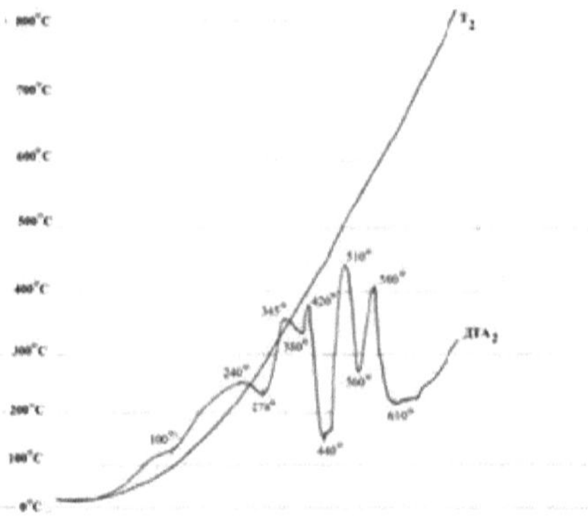

Fig.3.20. - DTA curve of chlorinated polyethylene

According to TG1 curve(Fig.3.18.) in the temperature range 20° C - 283° C no weight loss is observed, i.e. the polymer up to 283° C is thermally stable.

Then from 283° C the weight loss starts (breaking of weak bonds, endothermic effect on DTA 283° C) and in the temperature range 283° C-395° C the thermodestruction process takes place and at that the weight loss is 20%, i.e. the C-H bond chains in polyethylene are broken and they are snuffed out. And from 395° C-560° C the second stage of depolymerisation takes place and at the endothermic effect of 680° C the maximum weight loss is observed.

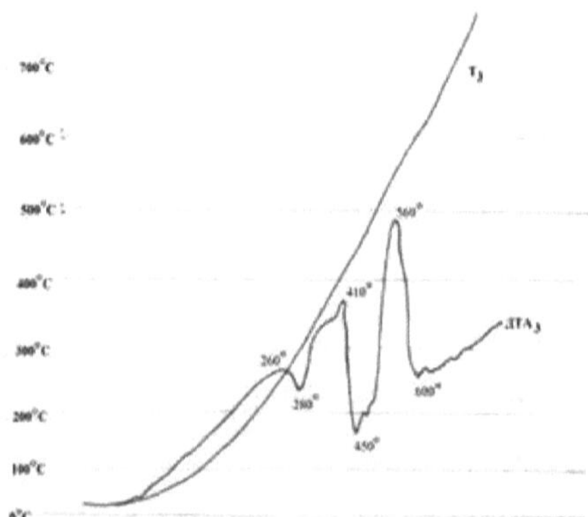

Fig.3.21.-DTA curve of CPE modified with DEA

The derivatogram of   chlorinated polyethylene (Fig. 3.20) shows the following endothermic effects at   temperatures 100° C,   278° C, 380° C,440° C, 560° C and 610° C . and the following exothermic effects - endothermic effects 240° C,345° C,420° C,510° C and 580° C. From the comparison of these data with Fig. 3.19, it can be concluded that the first endothermic effect of 100° C corresponds to the temperature of the beginning of softening of the polymer. Along with it on the curveDTG2

two new endothermic effects are found at 278° C and 380° C, accompanied with weight loss on the TG2 curve in the amount of 20% and 50%, which, apparently, is due to the breakage of weak bonds and the beginning of oxidative processes of

48

chlorinated polyethylene.

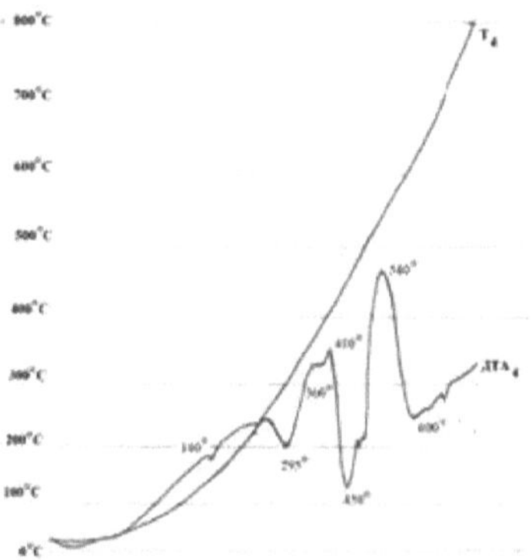

Fig.3.22. - DTA curve of CPE modified cTEA

This is confirmed by the exothermic effects on the DTA2 curve from 345° C and 420° C. The exothermic effect at 345° C corresponds to the first oxidation step of the chlorinated part of the polymer, and the exothermic effect at 420° C corresponds to the beginning of the oxidation process of the non-chlorinated part of polyethylene. At the exothermic effects at 510° C and 580° C, the second stage of the thermo-oxidative process takes place. At the endothermic effect of 560° C the depolymerisation of the chlorinated part of the polymer takes place and respectively from 580° C to 610° C the depolymerisation of the non-chlorinated part of the polyethylene takes place.

Comparing the curves of DTA3 (Fig.3.21.) and DTA4 (Fig.3.22.) with DTA1 (Fig.3.19) it can be seen that the endothermic peak at 120° C corresponding to the softening temperature of the polymer is almost absent on DTA3 and DTA4. This suggests that the modified chlorinated polyethylene exhibits a breakage of weak bonds contributing to these phenomena.

Comparing DTA3 and DTA4 we can consider similar indices of endothermic and exothermic effects, which indicates the flow of similar stages of thermo-oxidative degradation of the polymer, and whose indices are in accordance with literature data.

3.5. STUDY OF PHYSICAL AND MECHANICAL PROPERTIES OF THE OBTAINED OLIGOMERS

Table 3.3.

№	Sample names	Relative elongation at break, %	DTA data		
			Melting point, °C	Destruction temperature, °C	Temperature at which weight loss is 50%,° C
1	PE	320	120	283	410
2	Chlorinated PE	454	100	240	380
3	CPE+DEA	945		120	420
4	CPE+TEA	922	-	140	420

As can be seen from Table 3.3. chlorination of polyethylene and further amination of CPE with DEA and TEA is characterised by lower thermal stability compared to the initial polyethylene. Increasing the degree of functionalisation of these products causes a decrease in the onset temperature of its decomposition. Practical use of CPE and aminated CPE requires the use of stabilisers.

3.6. DETERMINATION OF DIFFUSION, SORPTION AND PERMEABILITY COEFFICIENTS OF THE CHEMICAL REAGENT THROUGH THE SAMPLES OF THE POLYMER UNDER STUDY

Several chemical reagents were selected for the test (Tab.3.4.):

Table 3.4.

№	Name of chemical reagent	Solution mass fraction, %	Note
1	Nitric acid	40	Obtained by dilution of concentrated nitric acid by GOST 701 with distilled water
2	Sodium carbonate	2	Obtained by dissolving sodium carbonate according to GOST 84 in distilled water
3	Sodium chloride	10	Obtained by dissolving sodium chloride according to GOST 4233 in distilled water
4	Sodium hydroxide	40	Obtained by dissolving sodium hydroxide according to GOST 4328 in distilled water
5	Sodium hydroxide	1	Obtained by dissolving sodium hydroxide according to GOST 4328 in distilled water
6	Sulphuric acid	98	Obtained by fortifying sulphuric acid according to GOST 4204 with oleum
7	Sulphuric acid	75	Obtained by diluting sulphuric acid according to GOST 4204 with distilled water
8	Sulphuric acid	30	Obtained by diluting sulphuric acid according to GOST 4204 with distilled water

During the tests, PE samples with diethanolamine and PE with thyrethanolamine were studied. Since these synthesised samples were obtained as a film (Fig.3.23.) within a day.

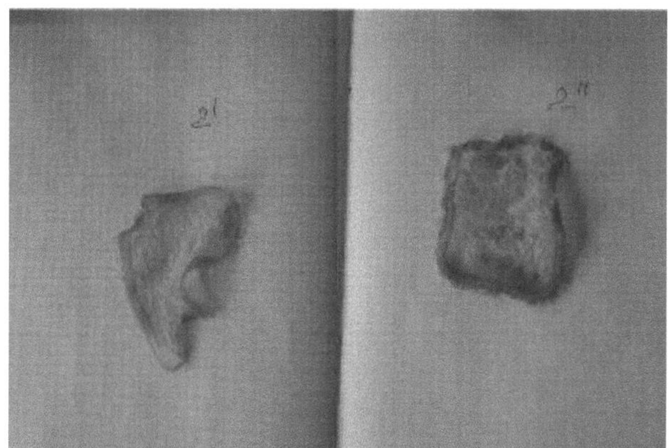

Fig.3.23. Samples of film obtained from CPE modified with DEA and TEA

For the chemical resistance test, the polymer samples were cut with a length of 2cm and a width of 1cm. The film thickness is approximately 0.1 cm.

Fig.3.24. Samples placed in vessels with chemical reagents

The samples were placed in vessels with chemical reagents (Figure 3.24).

Test temperature ±22° C.

The duration of testing of specimens were chosen according to GOST 12020-72 (8-16-24-48-96 h).

Sample of CPE polymer with diethanolamine

Table 3.5.

№	Name of chemical reagent	Time p intermediate measurements (hour)					
			8	16	24	48	96
		M (d)	M1	M2	M3	M4	M5
1	Nitric acid 40%	0,11	0,12	0,16	0,21	0,21	0,18
2	Sodium carbonate 2%	0,08	0,08	0,08	0,09	0,11	0,11
3	Sodium chloride 10%	0,06	0,06	0,06	0,06	0,07	0,07
4	Sodium hydroxide 40%	0,06	0,07	0,09	0,16	0,18	0,08
5	Sodium hydroxide 1%	0,15	0,15	0,15	0,16	0,16	0,17
6	Sulphuric acid 98%	0,12	0,13	0,14	0,18	0,19	0,12

| 7 | Sulphuric acid 75% | 0,14 | 0,14 | 0,15 | 0,19 | 0,2 | 0,15 |
| 8 | Sulphuric acid 30% | 0,15 | 0,15 | 0,15 | 0,15 | 0,16 | 0,16 |

Sample of CPE polymer with triethanolamine

Table3.6.

№	Name of chemical reagent		Intermediate measurement time (hour)				
			8	16	24	48	96
		M (d)	M1	M2	M3	M4	M5
1	Nitric acid 40%	0,16	0,18	0,2	0,3	0,27	0,26
2	Sodium carbonate 2%	0,11	0,12	0,15	0,2	0,21	0,18
3	Sodium chloride 10%	0,11	0,11	0,11	0,11	0,12	0,12
4	Sodium hydroxide 40%	0,1	0,13	0,15	0,2	0,24	0,13
5	Sodium hydroxide 1%	0,07	0,08	0,09	0,1	0,12	0,1
6	Sulphuric acid 98%	0,07	0,09	0,12	0,14	0,14	0,08
7	Sulphuric acid 75%	0,08	0,08	0,09	0,12	0,13	0,08
8	Sulphuric acid 30%	0,06	0,06	0,07	0,08	0,09	0,07

After the end of the specimen tests, the change in specimen weight after each test period was determined (ΔM) in per cent of weight gain using the formula:

$$\Delta M = \frac{(M_1 - M) \cdot 100}{M} \qquad (3.1)$$

№	Name of chemical reagent	Intermediate measurement time (hour)				
		8	16	24	48	96
		Change in specimen weight after each test period in per cent of weight gain				
		AM1	AM2	DM3	AM4	DM5
1	Nitric acid 40%	9,090909	45,45455	90,90909	90,909091	63,63636364
2	Sodium carbonate 2%	0	0	12,5	37,5	37,5
3	Sodium chloride 10%	0	0	0	16,666667	16,66666667
4	Sodium hydroxide 40%	16,66667	50	166,6667	200	33,33333333
5	Sodium hydroxide 1%	0	0	6,666667	6,6666667	13,33333333
6	Sulphuric acid 98%	8,333333	16,66667	50	58,333333	0
7	Sulphuric acid 75%	0	7,142857	35,71429	42,857143	7,142857143
8	Sulphuric acid 30%	0	0	0	6,6666667	6,666666667

Sample of CPE polymer with triethanolamine

Table 3.8.

№	Name of chemical reagent	Intermediate measurement time (hour)				
		8	16	24	48	96
		Change in specimen weight after each test period in per cent of weight gain				
		AM1	AM2	AMZ	AM4	DM5
1	Nitric acid 40%	12,5	25	87,5	68,75	62,5
2	Sodium carbonate 2%	9,090909	36,36364	81,81818	90,90909	63,63636
3	Sodium chloride 10%	0	0	0	9,090909	9,090909

4	Sodium hydroxide 40%	30	50	100	140	30
5	Sodium hydroxide 1%	14,28571	28,57143	42,85714	71,42857	42,85714
6	Sulphuric acid 98%	28,57143	71,42857	100	100	14,28571
7	Sulphuric acid 75%	0	12,5	50	62,5	0
8	Sulphuric acid 30%	0	16,66667	33,33333	50	16,66667

According to the obtained results, a graphical dependence is plotted ' $\Delta M = f(\tau)$:

Fig. 3.25. - Graph of change in mass of the polymer sample (chlorinated PE +Diethanolamine), characterising the sorption equilibrium and non-stability of the tested ones

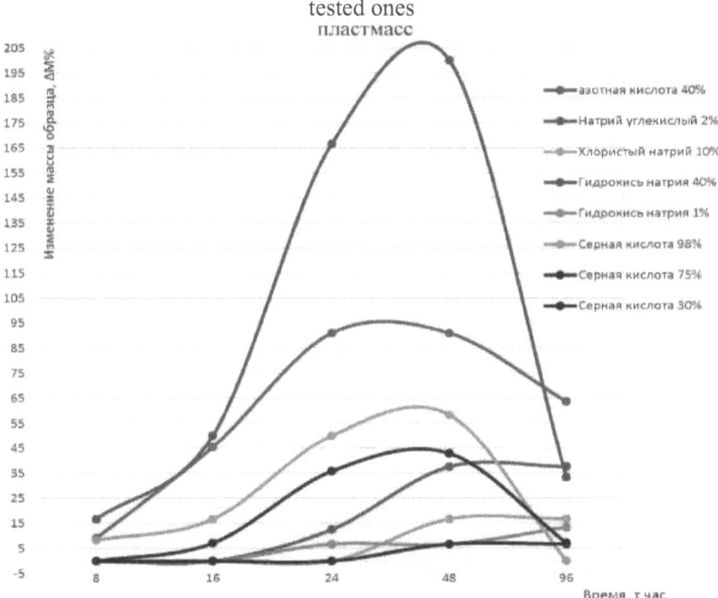

Fig.3.26. - Graph of change in mass of the polymer sample (chlorinated PE +Triethanolamine), characterising the sorption equilibrium and non-stability of the tested plastics

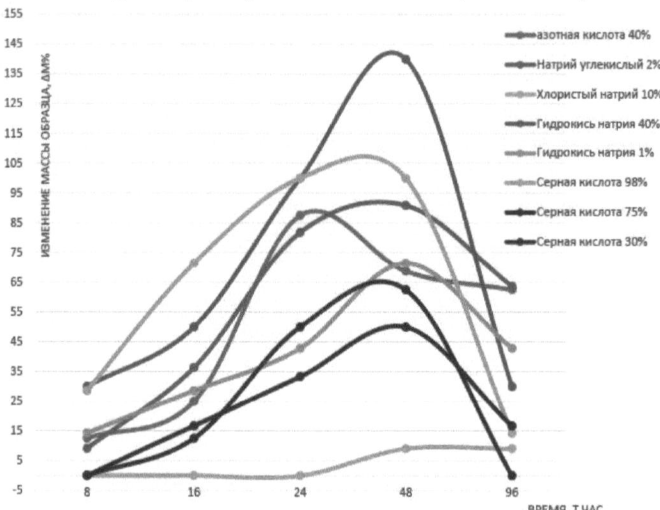

From the graph (Fig.3.25.-3.26.) we determined the time for which there was an increase in the mass of the sample to the value $M_{max}/2$, and calculated the diffusion coefficient of the chemical reagent in the polymer sample (D) in cm^2 /s by the formula:

$$Д = 0{,}0494\left(\frac{\tau_0}{\delta^2}\right)^{-1} \tag{3.2.}$$

Using the results we calculated the sorption coefficient of the chemical reagent in the polymer sample (S) in g/cm^3 , using the formula:

$$S = \frac{M_\rho}{V_{max}} \qquad M_\rho = M_{max} - M \tag{3.3.}$$

The permeability coefficient of the chemical reagent through the polymer samples (P) in g cm/cm^2 was calculated using the formula:

$$P = Д \cdot S \tag{3.4.}$$

Table 3.9.

№	Name of chemical reagent	sample thickness	time during which the mass of the sample increased	mass of chemical absorbed by the test sample	mass of the test sample at the established sorption equilibrium, g.	mass of the test sample before its first immersion in the chemical reagent, g	Diffusion coefficient of chemical reagent	sorption coefficient of chemical reagent in the sample polymer	Permeability coefficient of chemical reagent through samples
		δ, cm	τ_0 (sec)	Mr, g	Mshah, G	m, g	e (cm^2 /s)	S (g/cm)^3	P (cm2/s)
1	Nitric acid 40%	OD	58800	0,1	0,21	0,11	8,40* 10^{19}	0,5	4,20*10^{-9}
2	Sodium carbonate 2%	OD	108000	0,03	oh, and	0,08	4.57* IO^{19}	0,15	0,68*10^{-9}
3	Sodium chloride 10%	0,1	129600	0,01	0,07	0,06	3,81*10^{19}	0,05	0,19*10-9
4	Sodium hydroxide 40%	0,1	70200	0,12	0,18	0,06	7,03 *10^{19}	0,6	4,22*10-9
5	Sodium hydroxide 1%	0,1	72000	0,02	0,17	0,15	6,86* 10^{19}	0,1	0,68*10-9
6	Sulphuric acid 98%	0,1	68400	0,07	0,19	0,12	7,22* 10^{19}	0,35	2,53*10-9
7	Sulphuric acid 75%	0,1	72000	0,06	0,2	0,14	6,86* 10^{19}	oh,h	2,06*10-9
8	Sulphuric acid 30%	0,1	118800	0,01	0,16	0,15	4,16*10^{-9}	0,05	0,21*10-9

55

Table 3.10.

№	Name of chemical reagent	sample thickness	time during which the mass of the sample increased	mass of chemical reagent absorbed by the test subject	mass of the test sample at the established sorption equilibrium, g	mass of the test specimen before its first immersion in the	Diffusion coefficient of chemical reagent	sorption coefficient of chemical reagent in the polymer sample	Permeability coefficient of chemical reagent through samples
		δ, cm	τ_0 (sec)	Mr, g	Mtah, G	m, g	e (cm2/s)	S (g/cm3)	P (cm^2 /s)
1	Nitric acid 40%	OD	66600	0,11	0,27	0,16	$7,4*10^{'9}$	0,55	$4,08*10^{-9}$
2	Sodium carbonate 2%	OD	63000	0,1	0,21	oh, and	$7,8* 10^{'9}$	0,5	$3,9*10^{-9}$
3	Sodium chloride 10%	0,1	126000	0,01	0,12	0,11	$3,9*10^{'9}$	0,05	$1,9*10^{-10}$
4	Sodium hydroxide 40%	0,1	68400	0,14	0,24	0,1	$7,2*10^{'9}$	0,7	$5,05*10^{-9}$
5	Sodium hydroxide 1%	0,1	70200	0,05	0,12	0,07	$7,0*10^{'9}$	0,25	$1,76*10^{-9}$
6	Sulphuric acid 98%	0,1	43200	0,07	0,14	0,07	$0,11*10^{'9}$	0,35	$4,00*10^{-9}$
7	Sulphuric acid 75%	0,1	75600	0,05	0,13	0,08	$6,5*10^{'9}$	0,25	$1,63*10^{-9}$
8	Sulphuric acid 30%	0,1	75600	0,03	0,09	0,06	$6,5*10^{-9}$	0,15	$9,80*10^{-10}$

The change in the appearance of the samples was determined by visual comparison with the untested sample (Fig. 3.27). In this case, the changes in colour, gloss, presence of cracks, bubbles were determined.

It is recommended that the visual assessment of change in appearance be labelled as follows: O - no change, F - minor change, M - moderate measurement, L - significant change

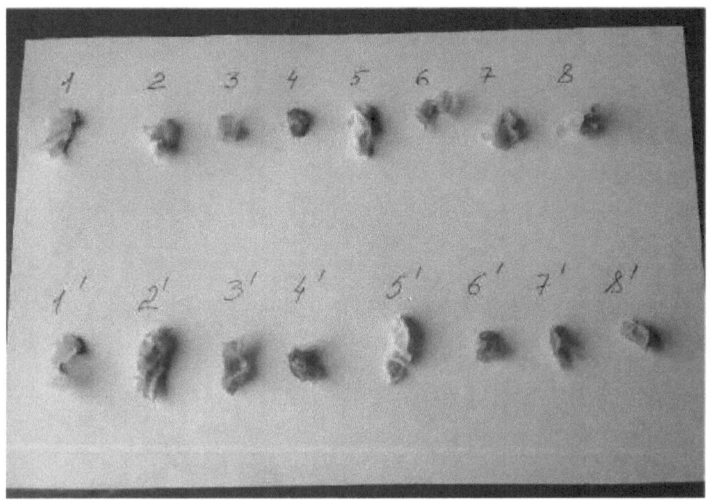

Fig.3.27. Appearance of specimens for visual comparison with the untested specimen.

Table 3.11.

№	Sample Name	Designation in the figure	Visual assessment of the change in the appearance of a CTE sample with	Designation in the figure	Visual assessment of the change in the appearance of a CTE sample with
1	Nitric acid 40%	1	L	1'	L
2	Sodium carbonate 2%	2	O	2'	F
3	Sodium chloride 10%	3	O	3'	O
4	Sodium hydroxide 40%	4	M	4'	M
5	Sodium hydroxide 1%	5	M	5'	M
6	Sulphuric acid 98 %	6	F	6'	F
7	Sulphuric acid 75%	7	F	7'	F
8	Sulphuric acid 30%	8	F	8'	F

3.7. ASSESSMENT OF POLYMER RESISTANCE TO CHEMICAL REAGENT EXPOSURE BASED ON CHANGES IN MECHANICAL PARAMETERS

Change of mechanical index (increase or decrease of its value compared to the initial value) after the samples stay in the reagent (ΔG) in per cent is calculated by the formula:

$$\Delta G = \frac{(G_1 - G) \cdot 100}{G}$$

$$(3.5.)$$

Where:

G - arithmetic mean value of the determined mechanical index in the initial state before immersion in the reagent;

G1 - arithmetic mean value of the determined mechanical index after soaking the samples in the reagent.

The results of the calculations are rounded to whole numbers.

Preliminary assessment of the resistance of plastics to the action of chemical reagent is carried out according to the change of mechanical parameters of polymers in accordance with GOST 12020-72.

Determination of changes in properties of polymer sample (CPE+DEA)

Table 3.12.

	Name of chemical reagent	G(r)	Intermediate measurement time (hour)					G 1
			8	16	24	48	96	
			Ml	M2	M3	M4	M5	
1	Nitric acid 40%	0,11	0,12	0,16	0,21	0,21	0,18	0,176
2	Sodium carbonate 2%	0,08	0,08	0,08	0,09	0,11	0,11	0,094
3	Sodium chloride 10%	0,06	0,06	0,06	0,06	0,07	0,07	0,064
4	Sodium hydroxide 40%	0,06	0,07	0,09	0,16	0,18	0,08	0,116
5	Sodium hydroxide 1%	0,15	0,15	0,15	0,16	0,16	0,17	0,158
6	Sulphuric acid 98%	0,12	0,13	0,14	0,18	0,19	0,12	0,152
7	Sulphuric acid 75%	0,14	0,14	0,15	0,19	0,2	0,15	0,166
8	Sulphuric acid 30%	0,15	0,15	0,15	0,15	0,16	0,16	0,154

Evaluation of resistance to the chemical reagent CPE + DEA

Table 3.13.

	Name of chemical reagent	G(r)	G1	AG	Resilience assessment
1	Nitric acid 40%	0,11	0,176	60	bad
2	Sodium carbonate 2%	0,08	0,094	17	satisfactory
3	Sodium chloride 10%	0,06	0,064	7	good
4	Sodium hydroxide 40%	0,06	0,116	93	bad
5	Sodium hydroxide 1%	0,15	0,158	5	good
6	Sulphuric acid 98%	0,12	0,152	27	satisfactory
7	Sulphuric acid 75%	0,14	0,166	18	satisfactory
8	Sulphuric acid 30%	0,15	0,154	2	good

Determination of changes in properties of polymer sample (CPE+TEA)

Table 3.14.

	Name of chemical reagent		Intermediate measurement time (hour)					
			8	16	24	48	96	

		G(r)	Ml	M2	M3	M4	M5	G 1
1	Nitric acid 40%	0,16	0,18	0,2	0,3	0,27	0,26	0,242
2	Sodium carbonate 2%	0,11	0,12	0,15	0,2	0,21	0,18	0,172
3	Sodium chloride 10%	0,11	0,11	0,11	0,11	0,12	0,12	0,114
4	Sodium hydroxide 40%	0,1	0,13	0,15	0,2	0,24	0,13	0,17
5	Sodium hydroxide 1%	0,07	0,08	0,09	0,1	0,12	0,1	0,098
6	Sulphuric acid 98%	0,07	0,09	0,12	0,14	0,14	0,08	0,114
7	Sulphuric acid 75%	0,08	0,08	0,09	0,12	0,13	0,08	0,1
8	Sulphuric acid 30%	0,06	0,06	0,07	0,08	0,09	0,07	0,074

Evaluation of resistance to the chemical reagent CPE + TEA   Table 3.15.

	Name of chemical reagent	G(r)	G1	AG	Resilience assessment
1	Nitric acid 40%	0,16	0,242	51	bad
2	Sodium carbonate 2%	0,11	0,172	56	bad
3	Sodium chloride 10%	0,11	0,114	4	good
4	Sodium hydroxide 40%	0,1	0,17	70	bad
5	Sodium hydroxide 1%	0,07	0,098	40	satisfactory
6	Sulphuric acid 98%	0,07	0,114	63	bad
7	Sulphuric acid 75%	0,08	0,1	25	satisfactory
8	Sulphuric acid 30%	0,06	0,074	23	satisfactory

3.8. SOLUBILITY STUDY OF THE OBTAINED OLIGOMERS

Chlorinated polyethylene is characterised by a large
chemically inert and resistant to most reactive substances, including acids and alkalis, at room temperature.

Chlorinated PE dissolves well in chlorinated hydrocarbons (carbon tetrachloride, chloroform, chlorobenzene, ditri- and tetrachloroethane, tetrachloroethylene, etc.) and aromatic hydrocarbons - benzene and its homologues, and cyclohexane. It is insoluble in water, alcohols, ketones, petroleum ether. The solubility of the modified polymer varies with chlorine content. Initially, with increasing chlorine content up to 30%, the solubility of CPE increases. At chlorine content of 54-56%, the polymer becomes poorly soluble. To eliminate the difficulties associated with the change in the solubility of CPE in carbon tetrachloride, its mixtures with other solvents with tetrachloroethane were used.

We have studied the solubility process of the synthesised oligomers in some solvents, including solubility processes in acids, alkalis and in water.

Table 3.16.

	PE	CTE	CPE+ DEA	CPE+ TEA	
Benzene	- - -	- - -	- - -	- - -	
Toluene	+ - -	+ - -	- - -	- - -	

	PE	CTE	CPE+ DEA	CPE+ TEA	
Dimethylformamide	- - -	- - -	- - -	- - -	
Heptan	- - -	- - -	- - -	- - -	
Water	- - -	- - -	- - -	- - -	
Alkali	- - -	+++	+++	+++	decomposes
Acid	- - -	- - -	- - -	- - -	
Ethyl acetate	- - -	- - -	- - -	- - -	

"- - - -" - absence of any changes in appearance, insoluble; "+ - - -" - very slight changes in appearance, slight swelling;

"+++++" - significant change in appearance, softening or collapse.

3.9. PROPOSED TECHNOLOGICAL PROCESS FOR PRODUCTION OF CHLORINATION AND SUBSEQUENT AMINATION OF POLYETHYLENE

Two methods are used to produce chlorinated polyethylene - gaseous chlorination of polymer solution in chlorinated solvent or chlorination of polymer suspension. The polymer is separated from the solution by distillation of the solvent with steam in the form of azeotropic mixture, from the suspension - by filtration or centrifugation.

Chlorination of polyethylene in solution consists of the following main steps:

1) Preparation of polyethylene solution;
2) Chlorination of polyethylene;
3) Stabilisation and separation of chlorinated polyethylene from solution;
4) Drying of the finished product.

The process of dissolution and production of chlorinated polyethylene from high-pressure polyethylene is carried out in reactor 1 - steel, enamelled apparatus with jacket and stirrer. The reactor has a barboter, through which chlorine is fed into it. In addition, it is equipped with a thermocouple sleeve, a sampling hatch and a loading hatch. A calculated amount of toluene is fed into reactor 1 and then polyethylene with the stirrer running.

Polyethylene is dissolved at 95-100° C for 2.5-3 hours. At the end of dissolution the first portion of initiator (porophor) is added and chlorination is started by feeding chlorine into reactor 1. Chlorination is carried out at 70-80° C. To maintain the set temperature in the jacket of reactor 1 is fed hot water.

Partially carried away toluene with acid gases goes to the reverse refrigerator 3, where its vapours condense, and through the phase separator 4 flow back to the reactor 1. Not condensed vapours go to the absorption column 5.

After 5.5-6 h of chlorination the gas supply is stopped. After the reaction mass has cooled down to 55-60° C, samples are taken to determine the chlorine content. In case of a positive result a modifier for amination (DEA or TEA) of chlorinated polyethylene is added to the reaction mass. Amination is carried out for 2-2,5 h with constant stirring, the temperature of the reaction mass is brought up to 90-100° C. The reaction mass is cooled down after the end of the process. After the end of the process the reaction mass is cooled to 40-50° C and the cooled solution of chlorinated and aminated polyethylene is purged with nitrogen to remove dissolved acid gases.

A stabiliser - 5% of ED-5 or ED-6 resin (from the polymer weight) - is introduced into the obtained chlorinated polyethylene solution, then the solution flows through the filter system 6 into the water extraction column 8. In the column 8 the solution is fed through a steam nozzle installed in its lower part. A neutral environment is maintained at the precipitation, which takes place in the aqueous phase.

The precipitated product in the form of fine crumbs is carried away by circulating water to the separator 10, where the vapour phase is separated from the liquid phase. Water vapour together with toluene vapour and fine particles of the product enters the trap 9. From the trap and separator the gas phase is fed to the phase separator 11, where condensation and separation of toluene and water takes place during cooling. From the separator 10, the circulating water with the product passes the hydraulic gate and enters the filter drum 12. Here the hot circulating water is separated from the product and the product is cooled sharply with cold water. Chlorinated polyethylene with quenching water goes to the vibrating screen 13, where the product is separated from water and further goes to the worm press 14 for drying.

From the phase separator 11, toluene is supplied for drying and purification by rectification, after which it is returned to the process.

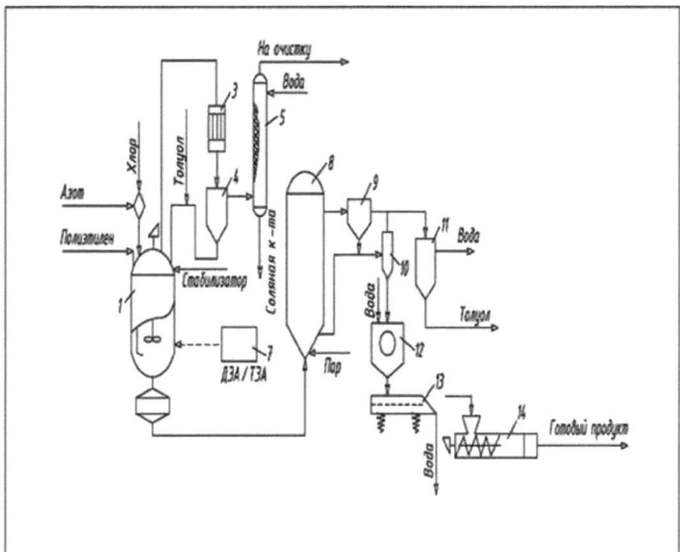

Fig.3.28. - Principal scheme of chlorinated polyethylene production:
1 - reactor, 2 - preheater, 3 - reverse refrigerator, 4,11 - phase separators, 5 - absorption column, 6 - filter, 7 - capacity for DEA or TEA, 8 - drop-off column, 9 - trap, 10 - separator, 12 - filter drum, 13 - vibrating screen, 14 - worm-press.

CONCLUSIONS

1.   New polymer halide- and amine-containing products based on polyethylene and its wastes were synthesised by direct halogenation followed by amination with DEA and TEA.
The influence of the degree of functionalisation of polyethylenes was studied by determining the diffusion, sorption and permeability coefficients of the chemical reagent, and the thermal stability, chemical resistance and mechanical properties of the obtained modified polymer products were evaluated.

2.   It is shown that chlorinated polyethylene and its wastes containing 25-30% of Cl_2 does not differ in structure from the initial polyethylene, retaining the properties of thermoplastic, but unlike polyethylene, exceeds the initial one in adhesion and elasticity.

3.   Products of interaction of CPE and its wastes with DEA and TEA have been investigated, the results of which have shown that the product with DEA is superior in chemical resistance.

4.   It is revealed that chlorination of polyethylene and further amination of CPE with DEA and TEA is characterised by lower thermal stability in comparison with initial polyethylene. Increasing the degree of functionalisation of these products causes a decrease in the onset temperature of its decomposition. Practical use of CPE and aminated CPE requires the use of stabilisers.

5.   Methods of chemical modification of polyethylene and its wastes and on their basis the proposed waste-free, resource-saving, low-tonnage technologies for production of chlorinated polymers are proposed.

LIST OF REFERENCES

1.   Report of the President of the Republic of Uzbekistan at the solemn meeting dedicated to the anniversary of the adoption of the Constitution of the Republic of Uzbekistan, 05.12.2016;

2.   Report by President of the Republic of Uzbekistan Islam Karimov at a meeting of the Cabinet of Ministers devoted to the results of socio-economic development of the country in 2014 and the most important priority areas of the economic programme for 2015;

3.   "UK leads the way in polymer waste recycling" Municipal Solid Waste, 2012, no.2. с.7.(РЖХим-13.05.-19.И.670.);

4.   http://www.polymery.ru/letter.php?cat_id=3 &n_id=1082;

5.   Asfandiyarov R.N. "Synthesis and properties of halogen derivatives of 1,2-polybutadiene", Dissertation, 2008;

6.   http://www.poliolefins.ru/ stat/polimer/1562-xlorirovannyj- poliyetilen.html;

7.   Nurkulov F.N., Jalilov A.T., Beknazarov H.S., Science yesterday, today, tomorrow: Collection of articles on the materials of 5 international scientific-practical conference, Novosibirsk, Oct.2013.No.5, p.1417. (RZhKhim 14.07.-19T.290. Investigation of physical and chemical properties of compositions based on chlorosulfated polyethylene);

8.   Gumarov A.H., Temkinova N.E., Rusanova S.N., Stoyanov O.V., Sofyina S.Y., Stroganov V.F., Mukhametova A.M., Gorinov R.M., Bulletin, Kazan. Technological University, 2014, 17, No.3, pp.117-123 (RZhKhim -14 .08.-19.T.270.   Materials on the basis of the chlorosulphated polyethylene);

9.   http://www.polymery.ru/letter.php?cat_id=3 &n_id=1082;

10. Timoshenko V.V., Tavroginskaya N.G., Composite materials. 2011.5,No.1,p.50-56, (RZhKhim - 13.02.-19.T.568. Development of polymer composites on the basis of PE waste of reduced flammability);

11. .http://http://www.polymery.ru/letter.php?cat_id=3&n_id=1082.polymery.ru/letter .php?cat_id=3&n_id=1082;

12.http://http://cementi.ru/jtextvvhvv198vvat.html.ru/jtextvvhvv198vvat.html;

13 . Otvalko J.A., Mironyuk V.P., Tverdov A.I., Kuzmin S.V. (n.s.), Trifonova E.V., Chlorinated polydiens with high content of vinyl links. Vulcanisates on their basis Journal "Rubber and Rubber", No. 2, 2012;

14 .Ingram I.S. "Improvement of industrial method of 3-chloro2-methylpropene production   and utilisation of by-products". products", Diss.work, Ufa, 2009;

15 .http://msd.com.ua/xlorirovannye-polimery/poluchenie-xlorirovannyx- polimerov-xlorirovanie-i-xlorsulfirovanie-poliolefinov/;

16 . A.A. Oshin, Industrial organochlorine products, 1984;

17 .Patent No. 2419635, Russia;

18 .Patent No. 471037, Canada;

19 Levin A.S. Chlorinated polyethylene and its application, 1973;

20 .Patent RU №2075483;

21 .Patent RU №2059650;

22 .Patent RU No. 2075483;

23 .SU Patent No. 753852;

24 .SU Patent No. 791250;

25 .SU Patent No. 415277;

26 .Patent RU №2148589;

27 .Patent RU №2170237;

28 .Patent RU №2170238;

29 .Patent RU No. 2186788;

30 .Patent of the Russian Federation No. 2497832;

31 .Patent RU No. 2217440;

32 .Patent RU №2252226;

33 .Patent RU No. 2244722;

34 .Patent RU №2337924;

35 .Patent RU №2046805;

36 .Patent RU №2429247;

37 .http://msd.com.ua/xlorirovannye-polimery/ xlorirovannye-polimery/;

38 .Patent No. 2928819, U.S.A;

39 . Jagatspanyan R.V., Mikosha V. I. - Reports of the Academy of Sciences, 1966, vol. 166, No.6, p.1405;

40 .Patent No. 2571901, U.S.A;

41 A.A. Goryacheva, "Kinetic regularities of modification of oligo- and polydienes by methods of chlorination and oxidative structuring", Diss.work. Tula, 2004;

42 .SU Patent No. 231108;

43 . Dontsov, A., Lozovik, G., Novitskaya, S., Chlorinated polymers, Moscow, 1979;

44 .KristonP, KoeskinaA, DimitrovM. KineticsandMechanizmsPolyreacts. Budapest, 1969, Preprs., v.5p.65;

45 KristonP, KoeskinaA, Dimitrov M. - J Appl. PolymerSei., 1970, v.14, No.11, p.2763;

46 .Patent of the Russian Federation No. 2469990;

47 . Dontsov A.A. et al. Rubber and Rubber, 1976, No.3, p.30;

48 . Timonin A.V. Research of anticorrosion compositions on the basis of chlorinated polyolefins in order to obtain coatings with increased adhesion, Diss.work, Moscow, 2001;

49 Andriasyan Yu.O. Elastomeric materials on the basis of rubbers subjected to mechanochemical haloid modification, Diss.work, Moscow, 2004;

50 Matveeva V.B. Polymer-textile material with increased protective properties, Diss.work, 2007;

51 .Patent No. 7388054, U.S.A. (RJChem - 09.24.-19.T.16.P.);

52 .Buznik V.M., Kiryukhin D.P., Nikitin A.N., III international scientific and

technical conference, Achievements of textile chemistry - in production, Abstracts of reports, Ivanovo, 2008, p.16-17, (RZhKhim - 09.19.-19.F.120);

53 .Ronkin G.M. New corrosion- and heat-resistant elastic polymeric materials for chemical productions / journal: Koks i khimiya. 2003. No. 5. pp. 32 - 34;

54 .Mulner M., Lozovik G.Y., Klinov I.Y., - Chem.prom.zarubezhnosti, 1971, No.10, p.70;

55 .Patent No. 3498934, U.S.A;

56 .Patent No. 3674597, U.S.A;

57 Ambrozashvili T.A., Krember M.L., Akutin M.S. - Plasticheskie massy. 1975, No.12, p.52;

58 Levin A.S. and others - Leather and footwear industry, 1975, №7, p.43;

59 .Patent No. 3658950, U.S.A;

60 Borisov D.N. Quaternary ammonium compounds on the basis of petrochemical raw materials: a-olefinovy   oxyethylated

nonylphenols, Diss.work, 2008;

61 .VNIISKA analysis of products of synthetic production

Kauchukov.Izdatelstvo "Khimiya" Moscow.1964. Leningrad.

Printed by Books on Demand GmbH, Norderstedt / Germany